Computational Analysis of Mathematical Systems and Wireless Networks

Computational Analysis of Mathematical Systems and Wireless Networks

Edited by
Dr. Ram Krishan
Dr. Rakesh Kumar
Dr. Sukhwinder Sharma
Dr. Anuj Kumar Sharma

CWP

Central West Publishing

Disclaimer
Every effort has been made by the publisher, editors and authors while preparing this book, however, no warranties are made regarding the accuracy and completeness of the content. The publisher, editors and authors disclaim without any limitation all warranties as well as any implied warranties about sales, along with fitness of the content for a particular purpose. Citation of any website and other information sources does not mean any endorsement from the publisher, editors and authors. For ascertaining the suitability of the contents contained herein for a particular lab or commercial use, consultation with the subject expert is needed. In addition, while using the information and methods contained herein, the practitioners and researchers need to be mindful for their own safety, along with the safety of others, including the professional parties and premises for whom they have professional responsibility. To the fullest extent of law, the publisher, editors and authors are not liable in all circumstances (special, incidental, and consequential) for any injury and/or damage to persons and property, along with any potential loss of profit and other commercial damages due to the use of any methods, products, guidelines, procedures contained in the material herein.

A catalogue record for this book is available from the National Library of Australia

NATIONAL LIBRARY OF AUSTRALIA

ISBN (print): 978-1-922617-48-4

Preface

We are pleased to present you the edited book titled "Computational Analysis of Mathematical Systems and Wireless Networks". In the present era, many methods for modeling and qualitative analysis of various kinds of mathematical and wireless systems are available. These analyzing methods show how to use predictive computational models to acquire and analyze knowledge about these systems. The book covers real-life examples of continuous computational scenarios. This book aims to capture new developments in the mathematical modeling of systems, as well as the performance computation of wireless networks. Our families and friends have been very supportive of the publication of this edited book. Finally, we are thankful to all who have contributed and spared their valuable time for this book.

Dr. Ram Krishan
Dr. Rakesh Kumar
Dr. Sukhwinder Sharma
Dr. Anuj Kumar Sharma

Table of Contents

CHAPTER 1

Estimation of Accurate Position of the Mobile Instrument in Wireless Sensor Network

Ambika N
Dept. of Computer Science and applications
St. Francis College, Bangalore, INDIA

Abstract: Sensors are tiny machines employed in multiple domains to track any object of interest. These gadgets are assembled to carry out the task. Instruments are static or dynamic. The mobile instrument is pre-programmed to drive from one stop to another. It will be assigned a particular task that it has to complete before reaching the pre-allotted location. Discovering the locale of the sensors is a problematic task. The proposed work aims to estimate the location giving 12% more accuracy than the previous work.

1.1 Introduction

Sensing an object of interest and providing readings are the major functionalities of the sensors[1]. These tiny gadgets[2] have received a good proportion of significance in the past few years. Minimizing human efforts these gadgets are used in many applications – home investigation [3][4], military scrutiny [5], habitat observation [6], etc. Sensors[7] are tiny gadgets used in many applications to track any object of interest. These instruments are assembled to accomplish the task. Machines are static or dynamic. The mobile instrument is pre-programmed to move from one end of the network to another. It will be assigned a particular task that it has to complete before reaching the pre-allotted location. Finding the location [8][9] of the sensors is a challenging task.

Three-dimensional positioning procedures based on indication strength gauge and time of arrival are proposed in [10]. Range-based localization methodology is used in the work. The distance is estimated between the unidentified instrument and the hosting instrument using this method. A dynamic hosting instrument is used in the recommended work. The instrument is assigned the responsibility to move on a pre-defined path. It broadcasts the position itself periodically. These positions act as virtual anchor instruments. Gauss-Markov mobility model is used in the work. The algorithm provides better flexibility, good coverage, and stability compared to other methodologies. The mobile instrument is time-dependent. The instrument is programmed to travel at a particular speed and direction. The angle between the speed and the coordinate axis is calculated in the work. The system practices the RSSI methodology to calculate the distance of the instrument. The attenuation value of the signal is determined from the logarithmic standard radio indication broadcast prototype. If the distance amongst the host instrument and the unidentified instrument is greater than the verge, the time-of-arrival methodology is utilized. An extreme possibility approximation is utilized to estimate the coordinate locale.

The proposed system uses auxiliary instruments in the network. These instruments are used as reference points in the system. The sensor instruments[11] use the position of the auxiliary instruments and the distance formula to derive the distance between them. Using the cosine rule, the angle of the intersection is calculated. Using the angle obtained and triangulation the distance between the detector instrument and the dynamic agent is estimated. The proposed work provides better accuracy by 12% compared to the previous contribution[10].

The recommendation is alienated into seven units. The overview is followed by the background in the second section. The literature survey is detailed in section 3. The planned system is detailed in section 4. Simulation done is NS2 is detailed in section 5. The future scope is conversed in unit 6. The contribution is concluded in section 7.

1.2 Background

The location of the instruments can be determined in many ways. One of them is the Angle of arrival measure[12]- w.r.t to known locations, the arrival time can be estimated. The location information of the calculating instrument and the reference instrument has to be accurate.

Triangulation utilizes the angle of received data and two reference points the triangulation is estimated. Let the network contain two references 1 & 2. Consider the object employed to evaluate the distance and the dynamic entity. Let the distance between the two references be d_1. Let d_2 be the distance between object O and R_2. Let d_3 be the distance between item O and R_1. Let the distance d_1 make an angle θ_1 with the distance d_2. Let the distance d_1 make an angle θ_2 with the distance d_3.

Let d be the distance between the O and M. then d is calculated by the formula-

$$d = d_1 \frac{\sin(\theta_1)\sin(\theta_2)}{\sin(\theta_1+\theta_2)} \tag{1}$$

1.3 Literature survey

Many contributions are made by various authors towards this field. The contributions are categorized into a homogeneous and heterogeneous system. The detailing in followed from the next section below.

Two location estimation techniques are suggested in [13]. The proposal tolerates a malicious attack against range-based location discovery. The least mean square evaluation is suggested by considering the observation. The malicious location references are considered in the observation. These malicious instruments behave inconsistently compared with benign ones. Using minimum mean square estimation methodology, the sensor location is assessed. The average square fault is estimated to formulate the degree of inconsistency. The voting-based location estimation is the second methodology proposed by the authors. The environment is split into a lattice of units. The proposal adopts the voting mechanism to reduce overhead and improve accuracy in the system. The target field is determining by collecting the reference location of the sensors. The minimum rectangular area is estimated considering the coverage of the location references. This procedure is followed by extending the area to find the beacon signal. Overlapping and iterative refinement is performed to measure the accurate location.

The position of a mobile device in an indoor environment is suggested in [14]. In the proposal, anchor instruments are used. These instruments aid in providing the sensor instruments with identification. The same is transmitted to the computer to evaluate the location of the sensor. The instruments residing in the vicinity of the anchors will receive the location of the anchor instruments. The computer is used to estimate the location of the anchor instruments by analyzing the received message from the sensors. The same is cross-verified with the data stored in the system. The system uses the database to estimate the sensor instrument location.

Secure localization is estimated in [15]. The two-tier network is considered in the study. The sensor instruments and reference points also known as locators are used in the proposal. The system aids the instruments to determine their position by broadcasting localization information. The origin of the message is verified and they serve as data selection points. Basic cryptographic primitives are used by the sensors and locators. Pair-wise keys[16] are obtained from the primary credential employing a pseudo-random function[17]. The suggested Robust Position Estimation Algorithm is used to determine the location and also verify the same. Using distance bounding with the sensor's verification is performed. The verification multi-lateration is applied to locate the sensors. The sensors estimate its location and notify the locators on the same or using distance bound. The distance bounding intersection region is performed. The methodology is designed to tackle wormhole[18] and Sybil attack[19].

A three-dimensional position procedure based on beacon strength gauge and time of coming is proposed in [10]. Range-based localization methodology is used in the work. The distance is estimated between the anonymous instrument and the anchor instrument using this method. A dynamic anchor instrument is used in the proposed work. The instrument is assigned the responsibility to move in a pre-defined path. It broadcasts the position of itself periodically. These positions act as virtual anchor instruments. Gauss-Markov mobility model is used in the work. The algorithm provides better flexibility, good coverage, and stability compared to other methodologies. The mobile instrument is time-dependent. The instrument is programmed to travel with a particular velocity and direction. The angle between the speed and the coordinate axis is calculated in the work. The system uses the RSSI methodology to calculate the distance of the instrument. The attenuation value of the signal is determined from the logarithmic standard radio indication communi-

4

cation prototype. If the distance between the host instrument and the unidentified instrument is greater than the edge, the time-of-arrival methodology is utilized. An extreme probability approximation is utilized to evaluate the coordinate position.

The location-based access control methodology is proposed in [20]. The processing delay in transmitting the packets and bandwidth limitations are considered in the work. The sensor instruments proceed with the Echo protocol. It uses radiofrequency[21] to transmit the nonce packet. The prover echoes the same to the verifier using ultrasound. The total time is estimated by the verifier. In the final iteration, multiple verification instruments are used. If the prover does not claim a position (region of acceptance) the execution is aborted. If the prover claims a position the Echo protocol procedure is followed.

The malicious instruments misleading the instruments with wrong information is detected and removed from the network in [22]. A unique pairwise key is shared between two communicating parties. The beacon instruments can identify the signals broadcasted by malicious instruments. The beacon instrument uses fake identification to request location information from the malicious beacon. Multiple detecting identifications are used by legitimate sign instruments to detect the guilty beacon instruments. To address mobile instruments maximum distance error is considered. The work also tackles the wormhole attack.

The radiofrequency methodology is used in [23] to evaluate the position of instruments in the wireless mesh network. The digital computer is connected to the instruments using the communication link. The identifiers are located at pre-defined locations. During instrument-to-instrument communication their identity of them is used. The same is transmitted to the system. The radio frequency signal used in the communication is estimated and the same is communicated to the computer. The angle of arrival and differential time of arrival is considered in the proposal.

The location of the mobile instruments is estimated in [24]. The procedure commences with the broadcast by the master sensor. It notifies the other instruments of the ultrasonic ranging measurement. This is followed by several transmission by the other instruments of the network. The sensors estimate the distance by measuring the acknowledgment. This measurement is shared with other instruments of the network in the vicinity. The information broadcasted by the mobile gadgets is used to determine the location of the same in [25]. The proposal uses triangulation methodology[26] from the sensors in the vicinity to estimate the position of the mobile device.

The location of the instruments is estimated in [27] in the presence of threats. A hybrid approach is suggested. The variation in antenna orientation and communication range is used to provide the estimation. The sensors collect information from the beacon over multiple transmissions. The range of intersection is estimated for all the transmissions. This estimation is compared with the previous estimation to obtain an accurate one. The proposal provides a good estimation of location in an environment with a wormhole attack and Sybil attack.

The location of the instruments is detected in [28] by using the theory of identifying codes[29]. The greedy algorithm[30] is used to make irreducible identifying codes. The network is considered as a graph made up of many vertices and edges. All the vertices are marked as unresolved. The identifying codes are applied for a vertex and made a comparison with other identifying codes applied to other vertices. This index is broadcasted if the comparison equals. If not resolved, the sets are updated.

Semi-definite programming is suggested in [31] to estimate the position in the environment. Quadratic formulas are utilized to generate to evaluate the location. Based on the equations, semi-definite programming models are generated. Euclidean distance[32] measure is used to approximate the distance between the two instruments. Reducing the sum of absolute errors is adopted in the proposal. The slack variables are introduced to minimize softer errors.

In [33] semi-definite programming methodology is proposed. A noisy environment is tackled in the work. The partial and imprecise distance statistics are used to arrive at the solution. The sensor's location is estimated using these distance constraints. Two symmetric matrices are considered. The identity of the matrices is found to estimate the 2-norm of the vector. To estimate the distance between the two sensors Euclidean distance is utilized. Semi-definite pro-gramming duality and interior-point algorithms are suggested in [34]. The symmetric matrix and identity matrix is used in the evalu-ation.

Location estimation of the sensors is done in [35]. The location of some of the deployed sensors is known and others are unknown. Received-signal strength[36] or time-of-arrival[37] is considered to estimate the location of the sensors. Cramer-Rao bound (CRB)[38] is estimated. The extreme-likelihood evaluators under Gaussian and log-normal representations are performed. These calculations pro-vide readings for time-of-entrance and received-beacon strength. Relative location estimation algorithms[39] are applied to arrive at the solution.

A statistical model for an indoor multichannel is proposed in [40]. The time of arrival and angle of coming is considered to evaluate the position. The data is set to move at a pace of 7 GHz. The time delay and amplitude are considered to calculate the mean square of arri-val data. The time arrival of data in the cluster is given using the Poisson process[41]. The angle of arrival is measured w.r.t the angle of the first cluster. Utilizing the estimated data, zero-mean Laplacian distribution[42] is estimated.

Cluster-based routing is adopted in [43]. The cluster instruments are deployed around the anchor instrument in the system. The in-struments cooperate in determining the position of the instruments with unknown locations. Some assumptions, checks and iterative refinements aid in providing a good estimation of the instruments. The instruments estimate the location using a least-mean square methodology where a instrument with an unknown position re-ceives range measurements from their neighbors. The triangula-tion[44] is solved to obtain an accurate position. In another scenario where the instrument receives range measurements from another, the assumption-based coordinates algorithm is used to find the site. The estimated place evidence is distributed to all the instruments of the network.

In [45] the work addresses the location discovery problem. Static instruments are considered in the work. The work uses some bea-con instruments that aid in estimating the location. These instruments broadcast signals to the network. Using the signals, the sensors will be able to estimate the angular bearings w.r.t the beacon instruments. The beacons are wired and hence are controlled. The beacons will be able to maintain identical angular speeds and achieve phase synchronization.

1.4 Working of the system

1.4.1 Representations employed in the study

Table 1.1. list of symbolizations employed in the contribution

Notations	Description
N	Network in consideration
N_i	i^{th} instrument of the system
$N_{i(x,y)}$	Location information of ith instrument of the system
Ack	Acknowledgement
Hello	Hello message
r	Communication range
A_i	i^{th} auxiliary instrument of the network
C_i	i^{th} group of the system
$A_{i(x,y)}$	Location information of ith auxiliary instrument of the system

1.4.2 Assumptions made in the study

- The grid is split into equal grids. The instruments are deployed manually.
- The auxiliary instruments are deployed at particular positions to guide the other instruments in estimating the location of the mobile agent[46]

- The mobile agent is capable of travelling from one end of the network to another. It transmits hello packets when it reaches a new position.

1.4.3 Types of instruments used in the study

The network consists of three types of instruments in the system-
- Stationary normal instruments- These instruments are positioned in the network to sense and transmit the processed readings to the pre-defined location.
- Auxiliary instruments[47] – These are the instruments that act as reference points in the network.
- Mobile instruments - The instruments are capable of moving from one point of the network to another performing various types of tasks assigned by the base station.

1.4.4   Cluster formation

After the instruments are deployed, they self-configure and communicate among themselves to form the cluster. The instrument broadcast Hello communication to the instruments in its neighborhood along with its location information. In representation (2) instrument N_i is propagating the Hello note to the grid N.

$$N_i \rightarrow Hello, L_i: N \tag{2}$$

The instruments which can listen to the message acknowledge themselves with the location information. In equation (3) instrument N_j is acknowledging to instrument N_i with an acknowledgement message Ack and location information L_j.

$$N_j \rightarrow Ack, L_j: N_i \tag{3}$$

After knowing each other, the instruments form a group and randomly select an instrument as its group head. The bunch leader is accountable to amass the information from its collection associates and onward the same to the following accessible hop or the base station.

1.4.5 Tracking the location of the mobile agent with the help of Auxiliary instruments

The mobile agent is programmed to move from one grid to another. The agents broadcast Hello messages when they position in a new position. In equation (4) mobile instrument M_i is distribution Hello note to the system N.

$$M_i \rightarrow Hello: N \tag{4}$$

The respective auxiliary instruments that fall on either side of the mobile agent broadcast their respective location readings to the cluster. Let A_i and A_j are the two auxiliary instruments that fall on either side of the mobile instrument M_i. In equations (5) and (6) A_i and A_j are broadcasting Hello message along with its location information to cluster C_i respectively.

$$A_i \rightarrow Hello, A_{i(x1,y1)}: C_i \tag{5}$$

$$A_j \rightarrow Hello, A_{j(x2,y2)}: C_i \tag{6}$$

1.4.6 Approximation of the position of the dynamic instrument

Utilizing the data sent by the auxiliary instruments in equation (1), the distance between the two auxiliary instruments is estimated. Let d_i be the respective distance between the auxiliary instruments. In equation (7) the distance d_i is estimated using location information $A_{i(x1,y1)}$ and $A_{j(x2,y2)}$. The coordinate's positions are used in the distance formula[48] to evaluate the distance between the two instruments.

$$d_i \rightarrow \sqrt{(x_2 - x_1)^2 + (y_2 - y_1)^2} \tag{7}$$

Considering the location point of the instruments and auxiliary instruments, the angle that is made is calculated. Let d_j be the distance between the detector instrument and auxiliary instrument A_i. In equation (8) d_j is calculated using location $A_{i(x1,y1)}$ and location of sensor instrument $N_{i(x3,y3)}$.

$$d_j \rightarrow \sqrt{(x_3 - x_1)^2 + (y_3 - y_1)^2} \tag{8}$$

Similarly in equation (9) d_k is calculated using location $A_{j(x2,y2)}$ and location of sensor instrument $N_{i(x3,y3)}$.

$$d_k \rightarrow \sqrt{(x_3 - x_2)^2 + (y_3 - y_2)^2} \qquad (9)$$

The angle of inclination is found using the cosine rule explained in [49]. θ_i and θ_j is calculated using the distance between the instruments in equations (10) and (11).

$$\theta_i \rightarrow \frac{\cos^{-1}(d_j^2 - d_k^2 - d_i^2)}{(2d_j d_k)} \qquad (10)$$

$$\theta_j \rightarrow \frac{\cos^{-1}(d_j^2 - d_k^2 - d_i^2)}{(2d_i d_k)} \qquad (11)$$

θ_i and θ_j is inserted in equation (12) to estimate the distance between the mobile instrument and the sensor is estimated.

$$d = d_i \frac{\sin(\theta_i)\sin(\theta_j)}{\sin(\theta_i + \theta_j)} \qquad (12)$$

1.5 Simulation

The contribution is imitated in NS2. Table 1.2 gives the list of various parameters used in the work.

Table 1.2 Parameters used in the simulation

Parameters used	Description
Area of surveillance	200m * 200m
Deployment of the instruments	manual
Number of sensors considered	90
Number of clusters	=(7 instruments * 10groups)+(5instruments * 4groups)
Number of auxiliary instruments	18

11

Time of simulation	60ms
Length of message (acknowledgement)	34 bits
Length of hello message	12 bits
Mobile instrument configuration	
Number of mobile agents considered	2
Maximum speed of transmission	115200 bps
Time interval w.r.t movement	32 s

1.5.1 Accuracy in estimating the position

The proposed work is compared with (Zhang, Yang, Zhang, & Yang, 2019). In the previous work, three-dimensional position procedure based on beacon strength indicator and time of arrival is suggested. Range-based localization methodology is used in the work. The distance is estimated between the anonymous instrument and the anchor instrument using this method. Gauss-Markov mobility model is used in the work. The mobile instrument is assigned the responsibility to move in a pre-defined path. It broadcasts the position of itself periodically. These positions act as virtual anchor instruments. The instrument is programmed to travel with a particular velocity and direction. The angle between the velocity and the coordinate axis is calculated in the work. The system uses the RSSI methodology to calculate the distance of the instrument. The attenuation value of the signal is determined from the logarithmic normal wireless signal transmission model. If the distance between the anchor instrument and the unknown instrument is greater than the threshold, the time-of-arrival methodology is utilized. The extreme likelihood evaluation is employed to approximate the coordinate location. The algorithm provides better flexibility, good coverage, and stability compared to other methodologies.

Three kinds of instruments are utilized in the suggested work. The static instruments are deployed in the network to sense and transmit the processed readings to the pre-defined location. The auxiliary instruments act as reference points in the network. Dynamic instruments are capable of moving from one point of the network to another performing various types of tasks assigned by the base station. The proposed work considers the known auxiliary instruments (Shao, Zhang, & Wang, 2014) as the reference points to evaluate the distance between the sensor instrument and the mobile agent. The distance formula is used to find the distance between the instruments followed by Triangulation. Utilizing this methodology, the instrument readings w.r.t to the mobile agent vary. Figure 1.1 provides the variation of readings between the cluster members. Figure 1.2 compares the proposed work with (Zhang, Yang, Zhang, & Yang, 2019) giving 12% more accuracy than the previous work. The system complexity of the proposal is O (n log n).

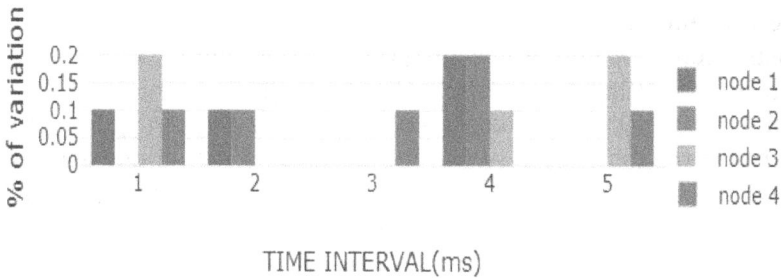

Figure 1.1. variation of readings between the cluster members

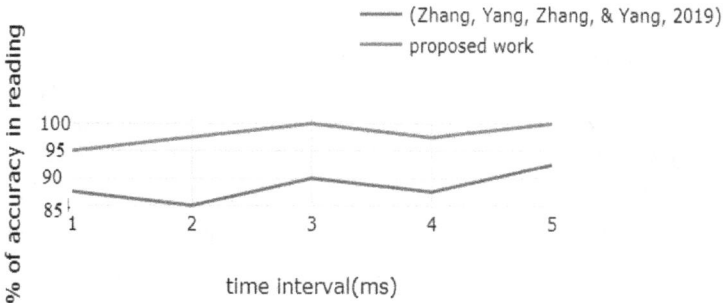

Figure 1.2. Comparison of work w.r.t accuracy

13

1.6 Future scope

The proposed work estimates the location of the mobile agent in different durations of time. The auxiliary instruments aid in the calculation. The instruments deployed can discover their location information and communicate the same. Based on the methodology adopted the readings of the sensors vary in the estimation of the location. Future work can focus on better methodologies to estimate accurate values w.r.t location.

1.7 Conclusion

Sensors are tiny gadgets that can self-configure and form a topology without human intervention. These networks come together to accomplish a task. The gadgets aid in minimizing human efforts. A Heterogeneous network is used in the proposed work. The auxiliary instruments located at specific points act as reference points. They aid in estimating the location of the mobile instrument at any instant of time. Two auxiliary instruments and a sensor join hands in deriving the distance between the sensor and the mobile agent. The new proposal provides better accuracy by 12% compared with the previous contribution.

References

[1]   Akyildiz, I. F., W. Su, Y. Sankarasubramaniam and E. Cayirci, "A survey on sensor networks," *IEEE communications magazine,* **40**(8), pp. 102-114, 2002.

[2]   Ambika.N, "Diffie-Hellman Algorithm Pedestal to Authenticate Nodes in Wireless Sensor Network.," in *Handbook of Wireless Sensor Networks: Issues and Challenges in Current Scenario's,* **1132**, Cham, Springer Nature, 2020, pp. 348-363.

[3]   N.K.Suryadevara and S. Mukhopadhyay, "Wireless sensor network based home monitoring system for wellness determination of elderly," *IEEE Sensors Journal,* pp. 1965-1972, 2012.

[4]   Â. Costa, J. Castillo, P. Novais, A. Fernández-Caballero and R. Simoes, "Sensor-driven agenda for intelligent home care of the elderly," *Expert Systems with Applications,* **39**(15), pp.12192-

12204, 2012.

[5] Lee, S. Hyuk, S. Lee, H. Song and H. S. Lee, "Wireless sensor network design for tactical military applications: Remote large-scale environments," in *IEEE Military communications conference(MILCOM)*, Boston, MA, USA, 2009.

[6] Mainwaring, Alan, D. Culler, J. Polastre, R. Szewczyk and J. Anderson, "Wireless sensor networks for habitat monitoring," in *1st ACM international workshop on Wireless sensor networks and applications*, Atlanta, Georgia, USA, 2002.

[7] C. S. Kubrusly and H. Malebranche, "Sensors and controllers location in distributed systems—A survey," *Automatica,* **21**(2), pp. 117-128., 1985.

[8] J. G. Wang, R. K. and S. K. Das, "A survey on sensor localization," *Journal of Control Theory and Applications,* **8**(1), pp. 2-11, 2010.

[9] P. Gajbhiye and A. Mahajan, " A survey of architecture and node deployment in wireless sensor network," in *First International Conference on the Applications of Digital Information and Web Technologies (ICADIWT)*, Ostrava, Czech Republic, 2008 .

[10] L. Zhang, Z. Yang, S. Zhang and H. Yang, "Three-Dimensional Localization Algorithm of WSN Nodes Based on RSSI-TOA and Single Mobile Anchor Node," *Journal of Electrical and computer Engineering,* pp. 1-8, 2019.

[11] A. Nagaraj, Introduction to Sensors in IoT and Cloud Computing Applications, UAE: Bentham Science Publishers., 2021.

[12] D. L. Fried, "Differential angle of arrival: theory, evaluation, and measurement feasibility," *Radio Science,* **10**(1), pp. 71-76, 1975.

[13] D. Liu, P. Ning and W. K. Du, "Attack-resistant location estimation in sensor networks," in *4th international symposium on Information processing in sensor networks*, Los Angeles, California, 2005.

[14] R. O. Farley, D. Kaleas and G. Giorgetti., "Sensor node positioning for location determination". U.S Patent 8,774,829, 8 July 2014.

[15] L. Loukas and R. Poovendran, "SeRLoc: Secure range-independent localization for wireless sensor networks," in *3rd ACM workshop on Wireless security.*, Philadelphia, PA, USA, 2004.

[16] Koblitz, Neal and A. Menezes, "Pairing-based cryptography at

high security levels," in *IMA International Conference on Cryptography and Coding*, 2005.

[17] R. Impagliazzo, L. A. Levin and M. Luby, "Pseudo-random generation from one-way functions," in *twenty-first annual ACM symposium on Theory of computing* , Seattle, Washington, USA, 1989.

[18] M. Bendjima and M. Feham, "Wormhole attack detection in wireless sensor networks," in *SAI Computing Conference (SAI), 2016*, London, UK, 2016.

[19] P. N. X. H. P. L. Mian Ahmad Jan, "A Sybil Attack Detection Scheme for a Centralized Clustering-based Hierarchical Network," in *IEEE Trustcom/BigDataSE/ISPA*, Helsinki, Finland, 2015.

[20] N. Sastry, U. Shankar and D. Wagner, "Secure verification of location claims," in *2nd ACM workshop on Wireless security*, San Diego, CA, USA , 2003.

[21] S. Bhaskar, "Is RFID technology secure and private?," in *RFID Handbook :Applications, technology, security and privacy*, 2008.

[22] D. Liu, P. Ning and W. Du, "Detecting Malicious Beacon Nodes for Secure Location Discovery in Wireless Sensor Networks," in *25th IEEE International Conference on Distributed Computing Systems (ICDCS'05)*, Columbus, OH, USA, 2005.

[23] J. McCoy, "Radio frequency location determination system and method with wireless mesh sensor networks". U.S. Patent 11/196,228, 15 March 2007.

[24] S. E. Adcook and C. D. Cook, "Relative location determination of mobile sensor nodes". US Patent US8416071B2.

[25] G. Cherian and H. Sampath, "Method for determining wireless device location based on proximate sensor devices". US Patent US 9,503,856 B2 , 22 november 2016.

[26] R. I. Hartley and P. Sturm, "Triangulation," *Computer vision and image understanding,* **68**(2), pp. 146-157., 1997.

[27] L. Lazos and R. Poovendran, "HiRLoc: high-resolution robust localization for wireless sensor networks," *IEEE Journal on selected areas in communications,* **24**(2), pp. 233-246, 2006.

[28] S. Ray, D. Starobinski, A. Trachtenberg and R. Ungrangsi, "Robust location detection with sensor networks," *IEEE Journal on Selected Areas in Communications,* **22**(6), pp. 1016-1025., 2004.

[29] U. Blass, I. Honkala and S. Litsyn, "Bounds on identifying codes," *Discrete Mathematics,* **241**(1-3), pp. 119-128., 2001.

[30] P. Slavık, "A tight analysis of the greedy algorithm for set cover.," *Journal of Algorithms,* **25**(2), pp. 237-254, 1997.

[31] P. &. Y. Y. Biswas, " Semidefinite programming for ad hoc wireless sensor network localization.," in *3rd international symposium on Information processing in sensor networks* , Berkeley, California, USA, 2004.

[32] P. E. Danielsson, "Euclidean distance mapping.," *Computer Graphics and image processing,* **14**(3), pp. 227-248, 1980.

[33] P. L. T. C. W. T. C. &. Y. Y. Biswas, "Semidefinite programming based algorithms for sensor network localization.," *ACM Transactions on Sensor Networks,* **2**, pp. 188-220., 2006.

[34] A. M.-C. &. Y. Y. So, " Theory of semidefinite programming for sensor network localization.," *Mathematical Programming,* **109**(2-3), pp. 367-384, 2007.

[35] N. Patwari, A. O. Hero, M. Perkins, N. S. Correal and R. J. O'dea, "Relative location estimation in wireless sensor networks.," *IEEE Transactions on signal processing,* **51**(8), pp. 2137-2148., 2003.

[36] C. Feng, W. S. A. Au, S. Valaee and Z. Tan, "Received-signal-strength-based indoor positioning using compressive sensing.," *IEEE Transactions on mobile computing,* **11**(12), pp. 1983-1993, 2011.

[37] Y. T. Chan, W. Y. Tsui, H. C. So and P. C. Ching, "Time-of-arrival based localization under NLOS conditions," *IEEE Transactions on Vehicular Technology,* **55**(1), pp. 17-24, 2006.

[38] P. Stoica and A. Nehorai, "MUSIC, maximum likelihood, and Cramer-Rao bound.," *IEEE Transactions on Acoustics, speech, and signal processing,* **37**(5), pp. 720-741, 1989.

[39] V. L. Boginski, C. W. Commander, P. M. Pardalos and Y. Ye, Sensors: theory, algorithms, and applications, **61**, Springer Science & Business Media., 2011.

[40] Q. H. Spencer, B. D. Jeffs, M. A. Jensen and A. L. Swindlehurst, "Modeling the statistical time and angle of arrival characteristics of an indoor multipath channel," *IEEE Journal on Selected areas in communications,* **18**(3), pp. 347-360, 2000.

[41] A. D. Barbour, L. Holst and S. Janson, Poisson approximation, **2**,

The Clarendon Press Oxford University Press., 1992.

[42] T. Eltoft, T. Kim and T. W. Lee, "On the multivariate Laplace distribution.," *IEEE Signal Processing Letters,* **13**(5), pp. 300-303, 2006.

[43] C. Savarese, J. M. Rabaey and J. Beutel, "Location in distributed ad-hoc wireless sensor networks," in *IEEE international conference on acoustics, speech, and signal processing.* , Salt Lake City, UT, USA, 2001.

[44] R. I. Hartley and P. Sturm, "Triangulation," *Computer vision and image understanding,* **68**(2), pp. 146-157, 1997.

[45] A. Nasipuri and K. Li, "A directionality based location discovery scheme for wireless sensor networks," in *1st ACM international workshop on Wireless sensor networks and applications*, Atlanta, Georgia, USA, 2002.

[46] H. Qi, Y. Xu and X. Wang, "Mobile-agent-based collaborative signal and information processing in sensor networks.," *Proceedings of the IEEE,* **91**(8), pp. 1172-1183., 2003.

[47] N. Ambika and G. T. Raju, "ECAWSN: eliminating compromised node with the help of auxiliary nodes in wireless sensor network.," *International Journal of Security and Networks,* **9**(2), pp. 78-84, 2014.

[48] E. Cheng, J. W. Grossman, L. Lipták, K. Qiu and Z. Shen, "Distance formula and shortest paths for the (n, k)-star graphs.," *Information Sciences,* **180**(9), pp. 1671-1680, 2010.

[49] K. Arai and R. A. Asmara, "3D Skeleton model derived from Kinect Depth Sensor Camera and its application to walking style quality evaluations," *International Journal of Advanced Research in Artificial Intelligence,* **2**(7), pp. 24-28, 2013.

CHAPTER 2

Security in WSNs: Attacks and Challenges

Navjot Sidhu and Monika Sachdeva
Department of Computer Science & Engineering, IKG Punjab
Technical University, Kapurthala

1.1 INTRODUCTION

Wireless Sensor Network (WSN) is the network of resource-restricted, small-sized, inexpensive and low-power sensor nodes. The small nodes comprise of sensing, communication and data processing components. These networks work on the combined efforts of the deployed tiny sensor nodes. The densely deployed sensor nodes monitor the phenomenon and send meaningful information to end-users. These sensor nodes are deployed to track ambient conditions. Aside from sensor nodes, base station and routers are other important components of WSNs. The base station serves as a bridge between the end-users and the sensor nodes. Routers are accountable for multi-hop communication within the network. Therefore, the sensor nodes send the data either directly to the base station or via one or more routers. These nodes are often deployed in unattended environments to observe and collect the required information. Due to many functionalities provided by these networks in terms of real-time applications, the demand for these networks is increasing day by day. However, there are multiple security issues in these networks due to which their performance can easily be degraded.

A brief introduction to WSNs and their available standards are given in this chapter. It also presents the security issues of these networks in terms of required security goals, classifications of attacks, and security challenges.

1.2 WIRELESS SENSOR NETWORKS

WSN is a special category of ad hoc networks that consist of a considerable number of small sensor nodes. These sensor nodes are equipped with multiple sensors, a radio interface, a power supply, a transceiver, an analog-to-digital converter, a memory and a processor. The different varieties of sensors can be attached to a sensor node to measure environmental conditions, such as temperature, light, vibration, sound, pollutants and pressure [1]. The sensor devices are spatially distributed in the monitoring field and work cooperatively to transmit information to the base station. The sensing nodes observe physical conditions of the environment. The observations are converted into digital signals by the analog-to-digital converter and then send to the processor. The processor with small storage capacity handles network monitoring tasks. Memory is used to store data and algorithms. A transceiver joins the sensor node to the network [2]. The power source commonly used is batteries. [3]–[6]. Fig. 1.1 shows the sensor node architecture.

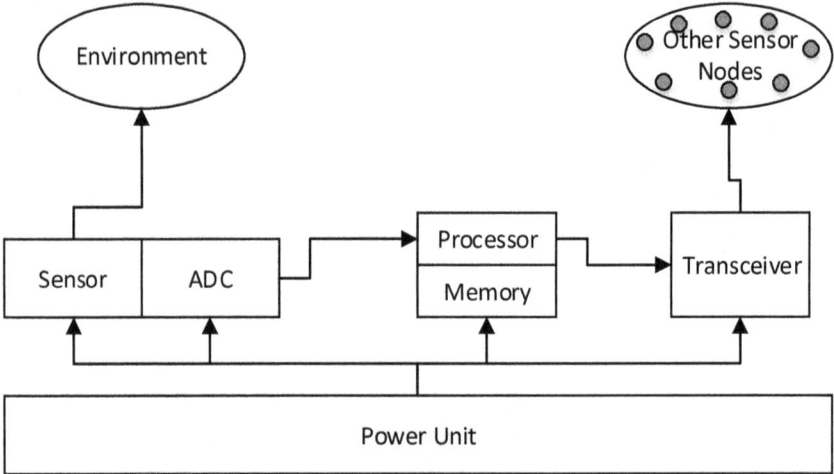

Figure 1.1 Sensor Node Architecture [2]

These networks have constrained resources including limited storage, communication range, battery power, processing capabilities and low bandwidth in each node. The network interference in the sensing environment reduces the connectivity between the sensor nodes [1].

The advantages of these networks, including the ability to measure the unattended environment, wireless capabilities, flexibility, scalability, accuracy, self-configuring, and relatively inexpensive make them suitable for diverse applications [4], [7].

1.2.1. WSN Standards

The hardware components of WSNs are chosen based on the requirements of the application for which these networks need to be deployed. A sensor node coordinates hardware and software for sensing, communication and data processing. It transmits and receives data from peer sensor nodes using wireless channels. The sensor node's lifetime depends considerably on the battery lifetime [2].

The less consumption of power is a key requirement for wireless sensor standards. A standard specifies the protocols and functions required for sensor nodes to operate and transmit data in the network successfully. Numerous standards are available in the market for WSNs such as IEEE 802.15.3, IEEE 802.15.4, ZigBee, IETF 6LoW-PAN and WirelessHART [1]. As every application has different requirements so based upon the application's requirement, the standard is used. For example, in some applications, extended battery life is required and some applications need secure data transmission [4].

1.2.2. WSN Applications

Sensor networks are used widely in multiple applications. More recently, interest has focused on security specific applications [2].

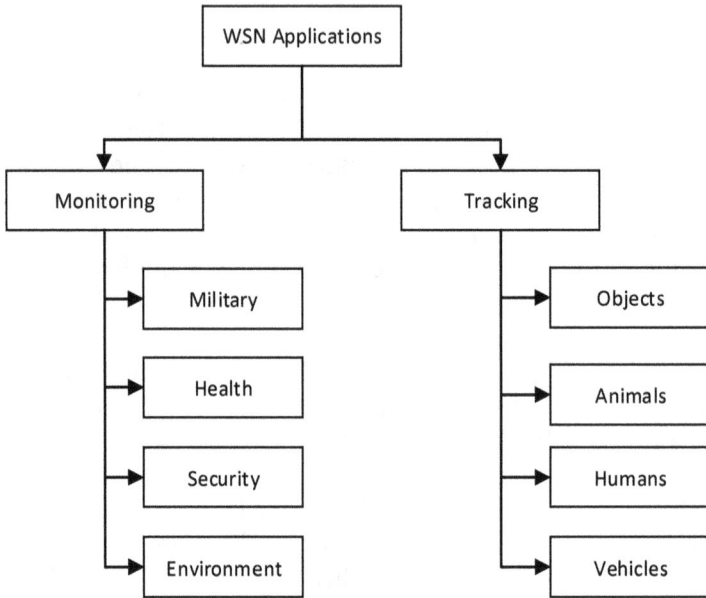

Figure 1.2 Classification of WSN Applications

The applications of WSNs can usually be divided into two categories (as shown in fig. 1.2): Tracking applications and monitoring applications. Monitoring applications include military forces monitoring, environmental monitoring, patient health monitoring, and vigilant surveillance. Tracking applications provide functionalities to track objects, animals, humans, and vehicles [1], [4], [8].

In short, in every aspect of today's life, a sensor is used. Common examples include detecting sound, vibrations, temperature level, check the water level, smart houses, smartphones, vehicles, and most importantly in the human body. A real-time WSN is therefore ubiquitously present in all facets of human beings [9]. Due to their wide variety of applications, WSNs have drawn a lot of attention from both industry and academia. However, various security threats are used by attackers, rendering the WSN systems weak and unstable. It is therefore essential to secure these sensor networks [10].

1.3. SECURITY IN WSN

As discussed in the previous section, WSNs are extensively used in many mission-critical applications like security, military and many

22

more. They deal with sensitive information. Hence, data from these networks is indeed quite important. However, the security in WSNs is much difficult to achieve than in traditional wireless and wired networks due to their resource limitations, open communication medium and deployment in hostile unattended environments. These networks actively monitor their surroundings and communicate wirelessly, which enable the chances of various active and passive attacks by an adversary effortlessly [2].

Even though traditional security mechanisms such as authentication and cryptography can protect up to some level, these solutions are not appropriate for resource constrained WSNs. When a sensor node is corrupted, it can launch multiple attacks like packet drop, packet modification, flooding the network with false data and Denial of Service (DoS) to degrade the whole network performance.
Due to the importance of these networks in daily life, these networks must require security to save their constrained resources from depletion and sensitive information from intruders. These networks must have security to attain security principles like availability, authentication, data freshness, confidentiality and integrity [10].

1.3.1. Terminology

The following key terms are used in this thesis:
1. Topology: Topology is defined as the deployment of sensor nodes in a monitoring area to transfer the information to base station [11].
2. Base Station: The base station, also called sink node is an interface between the sensor network and the user. It compile data from sensor nodes, infers useful information and forwards that to the user [11].
3. Attack, Intrusion: It is defined as an act to gain unauthorized access to network services, information and resources. The attacks are launched to compromise the sensor nodes in WSNs so that they could not perform their duties [12].
4. Attacker, Intruder, Adversary: These terms are used for a person who initiates malicious activities in the network to compromise network devices [12].
5. Vulnerability, flaw: This is the weakness of WSNs that can be exploited to launch malicious activities [12].

1.3.2. WSN Security Goals

The security in WSNs is at risk due to the use of open communication medium and specific vulnerabilities of these networks [11]. Briefly, the key security goal of a sensor network is to provide security to handle all types of attacks. [2], [10], [13] categorized the primary security services as follows:

- Authentication: Authentication is a process of proving that the originating node of a packet has created it. It allows a receiving sensor node to confirm that the data packet is sent by the specific sensor node. Therefore, in authentication, the source node proves its identity. It deals with identity confirmation of the participated sensor nodes and mainly differentiates malicious and legitimate sensor nodes in the network. In the case of WSNs, the base station and the sensor nodes must have capabilities to discriminate legitimate nodes from an attacker node. An attacker can compromise the legitimate sensor nodes and force compromised sensor devices to accept and forward fake data and routing packets. This false information causes unexpected network performance.

- Access-control: Access-control prevents unauthorized access to a network resources. The sensor nodes have limited resources to operate efficiently and perform their assigned tasks. These resources should not be assessed by any attacker node which consumes the available limited resources and halts the complete network operation.

- Confidentiality: Confidentiality protects the data packet's overall content in a message. A confidential message is that which should not be understood by anyone except from the desired message recipients. It ensures the protection of sensitive information from unauthorized users. Confidentiality protects the disclosure of sensitive sensor information while data transfer among sensor nodes and from sensor nodes to base station. The major problem is the existence of compromised nodes in the network, which are exploited by an attacker as these nodes can capture the critical information.

- Privacy: Privacy prevents adversaries to obtain private information. The private information such as the information of source node and available routes.

- Integrity: Integrity prevents information alteration during data transmission in the sensor networks. The altered information leads to major consequences. WSN applications such as environmental and healthcare monitoring rely completely on the integrity of the application therefore information should be protected.
- Authorization: Authorization ensures that only authorized sensor nodes can perform required operations in the sensor network, such as sending network status to base station after a specific time interval.
- Non-repudiation: It stops the source node from denying that it has transmitted a packet. Thus, with non-repudiation, a sensor node cannot refuse that the message has been sent by it, which it has actually sent.
- Freshness: Freshness ensures that a malicious node does not retransmit previously captured packets. It is highly recommended to save the limited network resources so that these resources could not be wasted to process the old information.
- Availability: It ensures the sustainability of network functionalities even in the situation of security attacks, mainly DoS attacks. It ensures that the users of a sensor network must obtain its services whenever required.
- Resilience: It is the network's capabilities to maintain the sensor network functionalities when some sensor nodes are destroyed and compromised.
- Forward and Backward Secrecy: Backward secrecy ensures that a newly joined sensor node could not listen to any message transmitted in the past. Forward secrecy ensures that a sensor could not listen to any network information after it has left the network.
- Survivability: It is the potential to operate with a minimum level of service in case of power loss, attacks and network failures.
- Self-organization: It is the capability of the sensor devices to reorganize and recover to work in any problematic situation.
- Auditing: It is the process of storing the information about each significant activity that happens inside the network.

1.3.3. WSN Attack Classes

The critical security goal in WSN is to save sensor devices from various attacks. It is required due to the inherent vulnerabilities of these networks due to which these networks can easily be compromised in several of ways, including eavesdropping, modification, fabrication and replication attacks [13]. To understand the details of the attack mechanism for each possible attack in WSNs, firstly it is required to understand different possible types of attack categories. [2] and [13] defined the following classes of attacks:

- Outsider Attacks: These are also called external attacks. These malicious acts are launched by sensor nodes, which are not the part of that WSN. In external attacks, an attacker adopts mechanisms to eavesdrop on secret information, injects false data and fabricates false data to disturb the regular operation of the entire sensor network.
- Insider Attacks: These attacks happen when legitimate nodes of a WSN behave abnormally. The adversaries enter the sensor network by breaking its security safeguards and compromise the sensor nodes to get crucial information from them. In internal attacks, the attackers compromise the sensor nodes to turn them into traitorous nodes.
- Passive Attacks: These attacks simply observe the traffic patterns such as eavesdropping or monitoring data packets exchanged within a WSN. It doesn't modify the data packets.
- Active Attacks: These attacks modify the data packets and transmit these packets in the sensor network. The attacker changes the important information about the network.
- Sensor-based Attacks: These attacks are launched by an adversary using sensor devices with similar abilities as the network sensor nodes.
- Laptop-based Attacks: These attacks are launched by an attacker using powerful devices such as a laptop computer. These devices have the highest energy, processing capabilities, and transmission range compared to the sensor network nodes. So they easily attract the sensor nodes by advertising false information and force other nodes to send their data through them to base station.

1.3.4. Attacks on WSN

Some attacks in WSN are identified in each ad hoc network, however, other attacks are specific to WSN. The attackers mostly exploit the vulnerabilities to attack these networks [11].

There are several reasons of the vulnerability in WSNs to network attacks. Firstly, the sensor nodes usually use the broadcast mechanism over an open transmission medium to inquire about their neighboring sensor nodes. Furthermore, these sensor nodes are deployed in a dangerous unattended environments where they are physically unprotected. For real-time sensor applications, it is not practical to protect each sensor node attacks and to observe these nodes physically. [2], [10]–[12] presented a detailed investigation on the possible attacks in WSNs. The adversaries can use one or more following mentioned attack methods to compromise WSNs.

1.   Eavesdropping

It is a passive attack that listens to the sensor network to hear transmitted information. An attacker observes network data by listening to the sensor nodes' communication. The attacker accesses the information easily when transmitted data is not encrypted. This attack does not cause any modifications to sensor data, therefore, it is difficult to detect.

2.   Radio Jamming

Radio jamming is an attack in which an attacker node sends wireless signals at a similar frequency that is used by the WSNs. The intention of the attacker is to congest the network so that the sensor nodes cannot transmit data. It is a kind of DoS attack in which the adversary disrupts the sensor network operation by continuously broadcasting radio signals.

3. Message Injection

This is an active attack that sends false messages in the sensor network. The goal of this attack is to send unnecessary information to modify the existing saved information and saturate the network resources.

4. Message Replication

This is another active attack, in which an attacker captures the packets transmitted on the network by some sensor nodes, and forwards them further to the wrong sensor nodes. The transmission of these fake data packets consumes the network as well as recipient sensor nodes' resources.

5. Node Destruction

It is a physical attack in which the sensor nodes are physically destroyed in WSNs. As these sensor nodes are placed in an unattended environment, which cannot always be observed. The attacker can swipe several sensor nodes, retrieve their information, and destroy them. The neighboring sensor nodes cannot operate if the link between two sensor nodes is destroyed by stealing intermediate sensor nodes. Moreover, the attacker reprograms and replaces the sensor nodes.

6. Denial of Service (DoS)

It is an active attack in which the attacker utilizes compromised sensor nodes and some high-power devices such as laptop computers to flood the network with a large number of useless messages. By flooding unwanted data in the network, this attack forces the sensor devices to remain active to perform complex computations and consume their battery power.

7. Hello Flood

This is an attack in which an attacker floods the sensor network with fake Hello messages which are transmitted using a high transmission power device like a laptop to compromise multiple sensor nodes of the network by announcing false neighbor status. The sensor nodes believe false routing information and transmit the data packets to the compromised nodes.

8. Black hole

It is an attack in which the attacker firstly deploys a malicious sensor node in the sensor network. This compromised sensor node sends the false information of routes to the source sensor nodes to force them to transmit their data packets through them. The sensor nodes forward their data packets towards these malicious nodes as they think that these nodes have the shortest path to the base station. These malicious sensor nodes do not forward the received data, and it results in a complete packet drop in the sensor network.

9. Gray hole

This is another packet dropping attack, in which an attacker spreads fake routing information in the sensor network through compromised sensor devices. When the neighboring sensor nodes send their data packets to these compromised devices, they drop some data packets resulted in selective packet drop. For this reason, it is also known as a selective forwarding attack.

10. Wormhole

This attack requires the participation of multiple malicious nodes in WSNs. The attacking sensor devices are connected by a powerful wired or wireless medium. These malicious nodes send the wrong route information in the sensor network and show that they have the quickest route towards the destination node. The sensor nodes send their information to these compromised devices.

11. Sinkhole

This is another selective packet drop attack, in which the compromised sensor node offers the quickest path to the neighboring sensor nodes. In this way, an attempt is made to divert traffic from the sensor network to pass through the compromised node. The neighboring sensor nodes transmit data for the base station to this malicious node. In this way, all information, which is required to transfer to the base station, capture by an attacker. Now, the attacker can launch another powerful attack by dropping important data packets.

12. Sybil

In this attack, the malicious sensor node pretends to be multiples sensor nodes. The compromised node modifies the routing table with incorrect information due to which the majority of sensor network traffic gets diverted towards the malicious sensor node.

13. Looping

It is an attack, in which an attacker compromises multiple sensor nodes of the network and instructs them to transmit an infinite number of false data packets in the network. It results in sensor nodes' energy depletion and saturates the network.

14. Message Alteration

In this active attack, the compromised sensor node captures a message and modifies it. It includes false information of the sender, receiver, alters data packets and also deletes some packets. In this way, it are corrupts important messages.

15. Slowdown

An attacker can slow down the network by using some malicious sensor nodes. It can use different attacks to perform network slowdown. Such that the compromised node sends multiple instructions in the sensor network to perform certain complex tasks. In this way, the attacker forces the victim node to remain active all the time, and it results in drainage of its battery.

16. Sensor specific

These are the malicious acts that attack the capabilities of deployed sensor nodes. It falsifies the output of any sensor node by physical means. In such a way, the attacker consumes the sensor network's limited resources.

As discussed in this section, different types of malicious acts are performed by intruders to compromise these networks. All the identified attacks affect the functions performed at each layer of the layered architecture of WSNs. To summarize the possibility of an attack on each architectural layer, table 1.1 shows the possible attack types against each layer of WSN layered architecture (as published in [14]). It also mentions the functions of each layer.

Table 1.1 Attacks on WSN layered Architecture

Layer	Function	Threat
Application	Use various application software	Selective Message Forwarding Data Aggregation Distortion
Transport	Maintain the flow of data Define communication connections	Flooding
Network	Provide routing information for communication	False Routing Injection Replication Black hole Sink hole Gray hole Wormhole DoS Sybil Hello Flood
Data Link	Provide power awareness and minimize collision	Packet Manipulation Spoofing Collision
Physical	Provide techniques, signal detection, modulation, frequency selection and encryption.	Eavesdropping Jamming Tempering

As discussed in [14] several DoS attacks can happen on the WSNs network layer. These attacks are classified into two categories, Routing attacks and Flooding attacks. Table 1.2 shows the attacks that can fall under these categories of attack classification.

1. Routing Attacks

These attacks use the routing protocols' vulnerabilities in WSNs. The attacker sends false routing packets to compromise the victim node. As a result of which the route formation for the victim node gets affected. The victim node in certain attack scenarios does not connect with the base station. After compromising the sensor nodes, attacker further launches packet forwarding attacks either complete or selective.

2. Flooding Attacks

The intention of an attacker can be consumption of victim nodes' resources. The resource may be a processing power, memory, or battery. So, flooding attack are also called resource consumption attack.

Table 1.2 Classification of Network Layer WSN Attacks

Attack Category	Attack Name
Flooding	Hello Flood RREQ Flood
Routing	Selective Forwarding Sinkhole Sybil Wormhole Black hole

These are the identified possible malicious acts performed by an attacker in WSNs. The adversaries disturb the normal operation of these networks by launching one or more attacks in the network. It is clear from the definition of each attack type that the attacker uses the network flaws to generate the malicious activities which degrade the network operation predominantly. To deal with these malicious activities, efficient solutions are required. However, to propose any effectual security solution, there is a need to comprehend the WSN security challenges.

1.3.5. WSN Security Challenges

The WSNs' security challenges need to recognize thoroughly to address the security problem. This section presents the WSN security challenges as discussed by [2], [6], [10], [11], [15].

- Ad hoc Deployment

Ad hoc placement of sensor nodes in these WSNs allows attackers to initiate several attacks ranging from active to passive.

- Dynamic Topology

Due to the limited battery power of tiny sensor nodes, node failure occurs in these networks. To keep connected with the base station, the sensor nodes keep on changing their topology. Some sensor nodes leave the network, and some new sensor nodes enter the network. The network should detect any error in the sensor nodes. For instance, sensor nodes are sensitive to a change in state, such as climatic change. In case it is not sending any data for a long period, it should verify itself that it is because of no change in the current state of a phenomenon or because the battery of the sensor node is low.

- Weak Security Protocols

Computationally complex security protocols, degrade the application performance in these networks. However, attackers can easily exploit simple security protocols. The existing complex security techniques that are used in other ad hoc networks are not appropriate for WSN because the sensor nodes have small battery capacity and computing power. The cryptographic solutions currently used for wired and wireless networks are not suitable for the limiting processors of sensor nodes. WSN needs effective routing protocols to secure data transmission in these networks.

- Open Communication

WSNs communicate using open wireless network channel. Any malicious node can enter the network to observe the channel transmission with a radio frequency configured at a similar band. Thus, attackers can easily listen to the network transmission in WSNs.

- Scalable

The scalability is one of the key characteristics of WSNs because sensor nodes need to be deployed in a large number to monitor any phenomenon. The sensor nodes in a network are deployed in the count of hundreds or even thousands. Therefore, security solutions also need to scale to large-scale deployments.

- Battery

Most of the sensor devices use battery. This battery power is always limited. Sensor nodes are deployed in a harsh environments, so these batteries are not replaced. Moreover, it is not feasible in these networks of thousands of sensor nodes, to identify a specific sensor which is battery deficient to replace its battery. Therefore, specific security protocols should be designed for sensor networks that use very little energy.

● Computation Power

The sensor nodes have low computation capabilities. Mostly the WSN applications need many sensor nodes. Therefore, these sensor nodes should be inexpensive. However, to minimize the cost of these devices, their computing power is also limiting. This low computing power does not effectively execute the complex security algorithms for sensor networks.

● Memory

The memory of the tiny sensor devices is very small, which is used to store specific application software in these nodes. Therefore, it is not feasible to execute complex secure algorithms for these networks.

● Transmission Range

The sensor nodes have a low transmission range. It is required to conserve battery capacity.

● Self- organization

The sensor nodes are placed in harmful hostile environments and it is not practical for the operators to repair faulty sensor nodes. Hence, the self-organization of sensor nodes is an essential requirement to form a connected sensor network.

There is always a conflict of interest in WSNs to minimize resource consumption and to maximize security. These key challenges need to examine carefully while proposing effective security solutions for WSNs.

1.4. CONCLUSION

As discussed in this chapter, recent advancements in WSNs technology allow the widespread use of these networks. WSNs play a significant role in daily life nowadays. Due to their use in large application areas, intruders are also continuously inventing new strategies to interrupt these networks. WSNs are vulnerable to many ad hoc network threats and WSN specific potential attacks. An attacker eavesdrops on information, compromises a sensor node, alters the data, injects fake information, and dissipates network resources. These sensor nodes broadcast their messages in the open medium. Therefore, the security issues of these networks need to address.

To accomplish security in WSNs, it is required to understand the requirement of specific security goals for WSN applications. For instance, confidentiality is not required in application areas such as environmental monitoring and home automation. But it is highly required for mission-critical applications, especially for military applications. There are several constraints to security solutions in WSNs such as storage, computation, processing and communication capabilities. These security challenges of WSNs need to keep in mind while proposing and testing new security solutions for these networks.

Some solutions are proposed by the fellow researchers to combat attacks in these networks. However, the inefficiency of the proposed methods to secure these networks in real time applications leads to the requirements of more appropriate solutions.

Funding: The work was fully financially supported by Department of Science & Technology (DST), New Delhi, India, under Women Scientist Scheme (WOS-A) with Grant Ref. No. SR/ WOS-A/ ET-1067/2014.

References

[1]   Yick, J., Mukherjee, B., and Ghosal, D. (2008) Wireless Sensor Network Survey. *Computer Networks*, **52**(12), 2292–2330.

[2]   Kavitha, T. and Sridharan, D. (2010) Security Vulnerabilities In Wireless Sensor Networks: A Survey. *Journal of Information Assurance and Security*, **5**(1), 031–044.

[3]   Pantazis, N. A., Nikolidakis, S. A., and Vergados, D. D. (2013) Energy-efficient routing protocols in wireless sensor networks: A survey. *IEEE Communications Surveys and Tutorials*, **15** (2), 551–591.

[4]   Rawat, P., Singh, K. D., Chaouchi, H., and Bonnin, J. M. (2014) Wireless Sensor Networks: A Survey on Recent Developments and Potential Synergies. *The Journal of Supercomputing*, **68**(1), 1–48.

[5]   Fahmy, H. M. A. (2016) Protocol Stack of WSNs. *Wireless Sensor Networks. Signals and Communication Technology.*,

Springer Singapore, 55–68.

[6]   Ishmanov, F. and Bin Zikria, Y. (2017) Trust Mechanisms to Secure Routing in Wireless Sensor Networks: Current State of the Research and Open Research Issues. *Journal of Sensors*, 2017, 1-16.

[7]   Guo, X. and Zhu, J. (2011) Research on security issues in Wireless Sensor Networks. *International Conference on Electronic and Mechanical Engineering and Information Technology*, 2, 636–639.

[8]   Kim, B. S., Park, H. S., Kim, K. H., Godfrey, D., and Il Kim, K. (2017) A survey on real-time communications in wireless sensor networks. *Wireless Communications and Mobile Computing*, 1–14.

[9]   Ali, A., Ming, Y., Chakraborty,S. and Iram, S. (2017) A Comprehensive Survey on Real-time Applications of WSN. *Future Internet*, **9**(4), 1–22.

[10]   Bhushan, B. and Sahoo, G. (2018) Recent advances in attacks, technical challenges, vulnerabilities and their countermeasures in wireless sensor networks. *Wireless Personal Communications*, **98**(2), 2037–2077.

[11]   Martins, D. and Guyennet, H. (2010) Wireless sensor network attacks and security mechanisms: A short survey. *13th International Conference on Network-Based Information Systems, NBiS 2010*, 313–320.

[12]   Wood, A. D. and Stankovic, J. A. (2004) A taxonomy for denial-of-service attacks in wireless sensor networks. *Handbook of Sensor Networks: Compact Wireless and Wired Sensing Systems*, 739–763.

[13]   Yu, Y., Li, K., Zhou, W., and Li, P. (2012) Trust mechanisms in wireless sensor networks: Attack analysis and countermeasures. *Journal of Network Computer Applications*, **35**(3), 867–880.

[14]   Sidhu, N. and Sachdeva, M. (2017) Taxonomy of Network layer DOS Attacks in Wireless Sensor Networks. *International Conference on Communication, Computing and Networking - 2017*, 1, 402–408.

[15]   Nagireddy, V. and Parwekar, P. (2019) Attacks in wireless sensor networks. *Smart Innovation, Systems and Technologies*, 105, 439–447.

CHAPTER 3

Securing SME Network Infrastructure with a Unified Threat Management System

Rimpi Rani
Computer Science, Punjabi University TPD Malwa College, Rampura Phul, India

Abstract:
Information systems and technology have become a vital element of many smaller and larger businesses' operations in recent decades, and have played a key part in dramatically altering and upgrading their company procedures on a regular basis. We wind up leaving and preserving confidential, important On computers, there is commercial and sensitive information as they become more and more integrated into our company organizations. Larger firms, on the whole, have the scientific competence and resources to provide more secure computer services. SME's, on the other hand, frequently lack the necessary platforms, infrastructure, technological expertise, and financial resources needed to use modern secure computing technology for the provision of computing services. This study reviewed the significance of network security, examined various types of network infrastructure threats, examined various approaches for mitigating network infrastructure vulnerabilities, and offered a way for safeguarding the network infrastructure of small and medium-sized businesses. as the network's first line of defence, depends on the connections between each tier of the distribution switch, which provides zone-based surveillance and regulating mechanism to safeguard any risks to the network.

Introduction: Viruses and spyware, among other things, are constantly threatening our computers and networks, posing a threat to our files data , operating systems, as well as the entire network. Many businesses overlook the importance of security. Company's network, the most crucial aspects of any company with privileged knowledge [1]. Security on the Internet and LANs is becoming one of the most pressing concerns for any company. Information and network security threats have increased and spread considerably as networking and the Internet have evolved. Many of these risks have evolved into actual attacks that cause damage, theft, or intellectual data leaks. As a result, both businesses and governments have become increasingly aware of the need of network security. As a result of this awareness, experts have begun to look for the source of these risks. To prevent this, there are instruments and strategies available. Despite the fact that there are security dangers, only with the right tools can you combat them. The technology and approaches will be implemented successfully. In the basic network architecture of a Small and Medium Enterprises (SME), this article examines the relevance of network security, potential threats, and techniques to reducing network security threats both inbound and outgoing.

1. **Network Security's Importance**: Local area networks (LANs) in organizations, there is a vast amount of data and it is expanding. The majority of information in most firms is confidential, including client information, Important paperwork, company strategies, asset reports, and more all available. DDoS attacks on Visa and MasterCard websites have recently occurred after they severed Wikileaks' services; the downtime created by such attacks has resulted in serious financial losses for the corporations [2]. Once the organization's network security is breached, hackers may gain access to clients' identification cards, bank accounts, and other private information in order to use it for their gain. It is the organization's sole responsibility to safeguard confidential information, such as that of the persons with whom you do business. With effective network security, it is possible to prevent unauthorized access to data on networked computers, as well as maintain them safe and secure from any dangers. Viruses, spyware, malware and worms are examples of threats. These modified applications, scripts, and background processes can cause major computer harm by destroying files, modifying data in files, transferring information contained in files to others who can read it, and so on, change computer settings, and eventually ruin a computer to the point of irreversible destruction [3]. Clicking on a poisoned web link, reading a tainted email link, and accepting, among other things, tracking devices such as cookies, activate all of the previously mentioned security flaws. To assist prevent These types of events necessitate the implementation of security measures that evaluate all recommended it are all examples of precautionary steps.

2 Type of Networks: A network's classification is mostly deter-mined by the extent of the network's coverage region shown in fig-ure1.

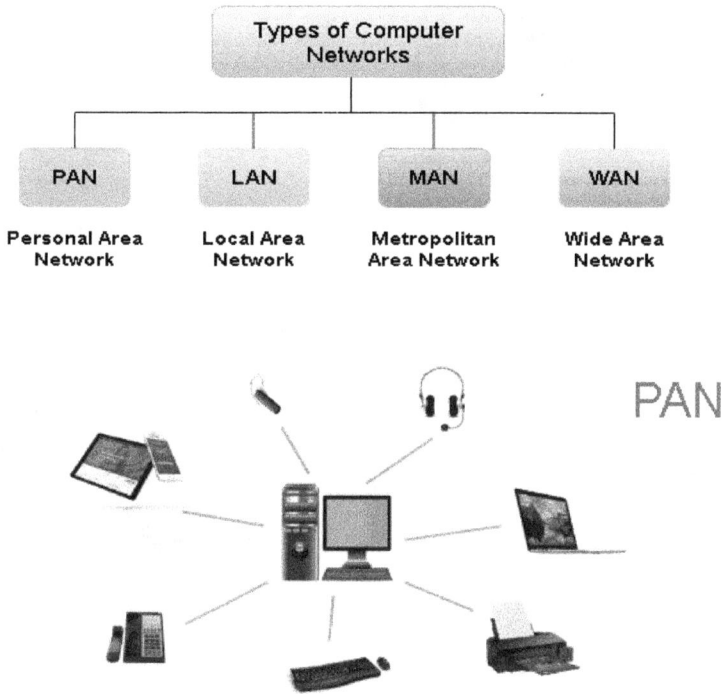

```
                    Types of Computer
                        Networks

        PAN          LAN          MAN          WAN

   Personal Area   Local Area   Metropolitan   Wide Area
      Network       Network     Area Network    Network
```

PAN

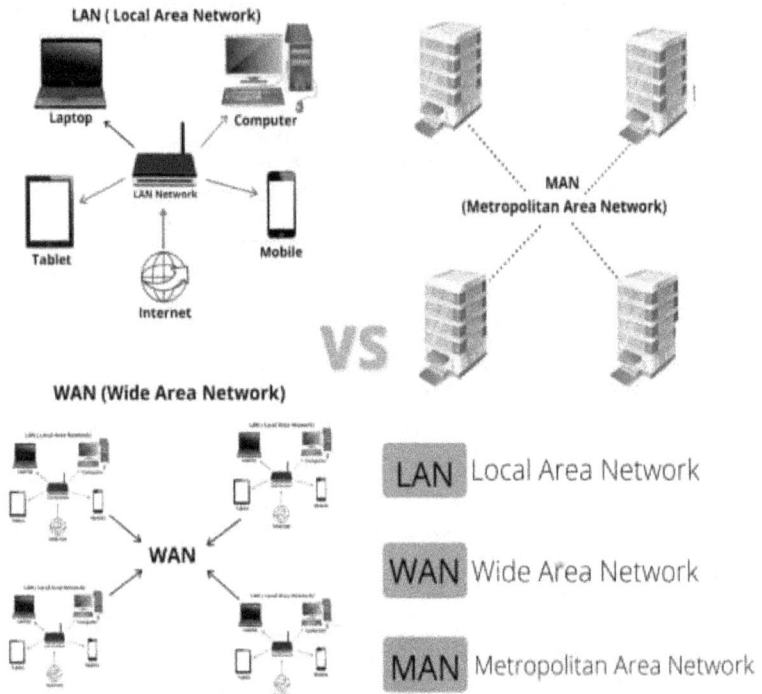

Figure 1. Types of Networks

PANs are a type of small network that typically span only a few meters and are used to communicate between computing devices like PDAs and mobile phones LANs (Local Area Networks) are networks that serve a specific geographic area that cover a specific geographic area, They are generally used for communication of network and can be found in places like a home or an office building [4]. Wireless networks that cover a whole city are known as Metropolitan Area Networks (MANs) to link a variety of computers together, such as ATMs as well as private organization spread out over a vast region. Finally, Wide Region Networks (WANs) are networks that span a huge area of land, often even crossing state and national borders, and are mostly used to connect organizations in various places of the globe [4].

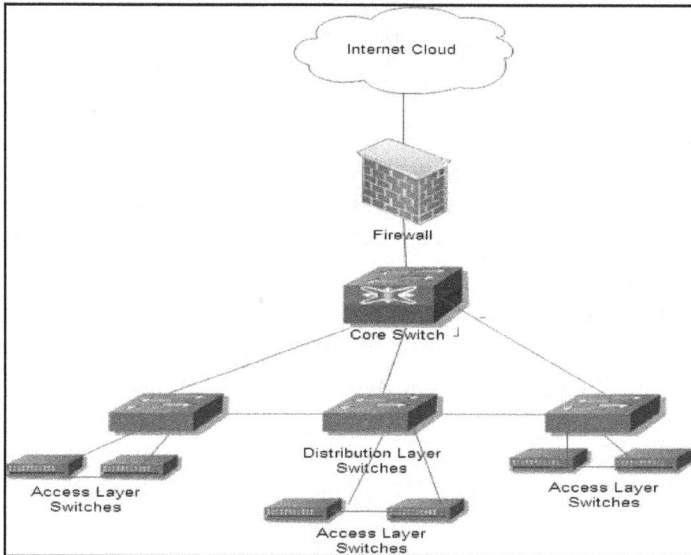

Figure 2. Network schematic of the general organization

3. Threats to the Network: A common communication consists of lots of divisions, any of which can be vulnerable to illegal activity. We look at some of the ways network security might be improved and can be jeopardized. Professional hackers and company rivals can all compromise routers, switches, and hosts. Internal network attacks are taken into account. To find the most effective techniques for defending against attacks, To begin, we must first understood the different types of illegal that might be undertaken, as well as the data harm these illegal do it[1]. DoS attacks are especially harmful but, despite the fact that they don't provide outsiders access to specific information, they can wreak a lot of damage, They "lockdown" its resources, making it impossible for legitimate users to access them. Based on Jerry Fitz Gerald's basic architecture offered [4], Figure 1 displays a typical SME network design and infrastructure. Within the constraints of LAN accessing applications, we will assess the threats and describe the technology and strategies that can be used to minimize them. These attacks are usually carried out by sending massive amounts of jumbled or otherwise unmanageable data to machines connected to business networks or the Internet. DoS (Distributed Denial of Service) assaults are even more dangerous. An attacker gains control of a number of computers or hosts. Just as to a report from 2002 assessment by the Computer Security

41

Institute (CSI) 38, as well as the FBI's "Computer Crime and Security Survey." DoS assaults were detected by % of respondents, compared to In the year 2000, the figure was 11%. Moore et al [5] claim that They use updated backscatter analysis in their operations and Every week, there are roughly 2050-3050 active DoS attacks. Over the course of three years, Researchers found 67,800 attacks on over 35,800 different Internet servers belonging to over 5,400 different businesses [5]. The recent (DDoS) attacks on Visa and Master Card, which knocked both services offline, are a good example of this, underscore the significance of such attacks, which can result in significant financial damage [2]. Historically, the most common type of assault has been password attacks, In order to gain unlawful access to confidential information, a culprit obtains unauthorized access to network passwords. If hacker "cracks" He uses a legal user's password to obtain access to that user's network resources. as well as a normally extremely powerful platform access to network. For example, in December 2010, a hacker at Seattle's University of china Centre stole user credentials and access to personal data files, about around 5020 patients [6]. Because people typically use common phrases or numbers as passwords, hackers can easily obtain them allowing them to employ software tools to painstakingly identify those passwords. Social engineering tactics are also used by hackers to acquire access to passwords [7]. The act of using non-technical methods to gather confidential network security information, such as providing technical support and making direct phone calls to employees in order to obtain information about the password, is known as social engineering. Since the Internet's early years, when the network consisted solely of email servers, The ultimate goal of the hackers was to get root access to the targets [8]. With root access, the hacker had complete control over the system, and he could frequently gather enough data. Hackers have been more interested in e-business application hosting. Security flaws, or security holes, in these hosts' have neglected to safeguard are routinely exploited by hackers. Hackers obtain additional assaults. E-mail is another sort of danger that can disrupt corporate networks. Spam creates managerial and technical problems. aspect. Exploring spam emails and making phone calls is a waste of time. Ensure that no important business emails are flagged as garbage. SPAM squandered bandwidth and storage space on a technical level. The company's information is protected servers by grabbing and storing these emails These strategies might result in substantial financial losses as well as legal consequences. Organizations suffer several hours or days of an outage as a result of network attacks, as well as major data confidentiality and integrity breaches[8].

4. Security Network Methodologies

Firewalls: This is the initial line of security in what should be a multi-level computer defense system. A firewall is software that monitors and controls network traffic both inbound and outbound whether or whether certain types of traffic should be prohibited or provided by the user [9]. The word "firewall" has come to refer to a device or group of devices that "protects its occupants from potentially dangerous external environments," such as the Internet [10]. A policy is a set of guidelines. The routers used to segregate networks in the late 1980s were the forerunners to firewalls for network security [10]. There was no widely used internet at this period in history; Behind that router, There was only the local network, which you were a part of. The way individuals communicate has changed since the internet's introduction, new dangers appeared that would quickly exploit the susceptibility of network security flaws [10]. Firewall groups are also available. Packet filtering firewalls look at packets, the fundamental units of data transfer and those that comply with the policymaker's rules are sorted out. Users can build rules that allow them to open ports and allow particular IP addresses to access their systems, or, to run. Stateful and stateless firewalls are the two subtypes of this sort of firewall. Stateful firewalls keep track of current sessions and exploit this "state information" to improve packet processing speed [11]. Several attributes are characterized using parameters such as source and destination IP addresses, UD ports, and so on. Stateless firewalls have no way of knowing where they are in the filtering process. As a result, it is unable to make sophisticated, In certain instances, decisions must be made [12]. Circuit gateways keep track of the connections between two machines. Rather than inspecting each just checks to determine if it is valid if the connection session request is valid. They are unable to examine data transmission between networks (in general), however, they can (manage) direct network connections, similar to filtering gateways. The application gateway, a new kind of firewall, is the final option (Layer 7 Firewalls). This sort of firewall performs a variety of functions, including blocking undesired The Firewall administrator can specify URLs and file extensions. Modern Layer 7 firewalls are incredibly adaptable. What kind of internet access do they

43

have? [13]. They are also capable of doing deep packet inspection, which compares each packet against a signature in the firewall database to guarantee that the packets are safe. Enabling ingress and egress on the firewall's WAN link eliminates the possibility of IP spoofing attacks [13]. Computer security relies heavily on firewalls. They're on the front lines of the fight against harmful software [10]. Because of the human defects that exist in every programmer and operating system, it is vital to have a firewall to help guard against hostile programmers that aim to exploit these weaknesses.

5. Security Services for Gateways in Their Entirety (CGSS): Anti-Virus, Anti-Spyware, and Intrusion Prevention Service at Gateway, and Application Intelligence Service is a service that provides intelligent information about applications., for preventing data leaks and offering detailed data analysis. [14] Controls at the application level Such a service exists. combines all types of security into a single package Unified Threat Management is a common name for this device [14] (UTM). Larger enterprises with larger throughput needed specialized appliances. The idea is the same for each module, but the on-demand requirements differ based on the business. The UTM cleans and checks the integrity: Figure 2 depicts how the UTM operates.

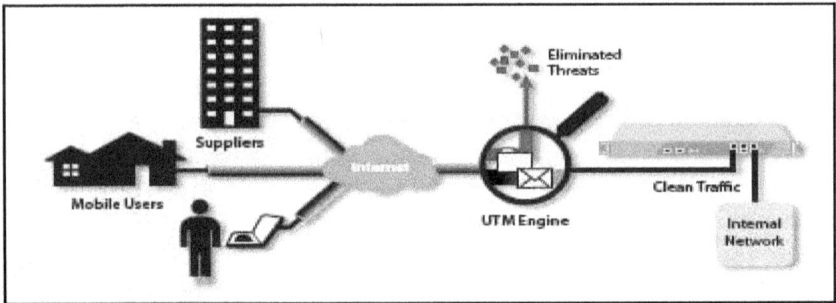

Figure 3. (adapted from SonicWALL, 2007): UTM Overview

Figure 3 depicts the design of the CGSS dynamic Sonicwall-based security model technologies. The CGSS modal makes it very apparent that getting The vendor's most recent signatures and updates are critical for keeping the network secure against the most recent attacks and vulnerabilities. The Unified Threat Management Solution will assess threats as shown in Figure 3. Traffic on the various

OSI network layers and make certain packets being transmitted in the in good shape; further information on the Sonicwall UTM architecture [14] can be used to refer to inspection methodologies.

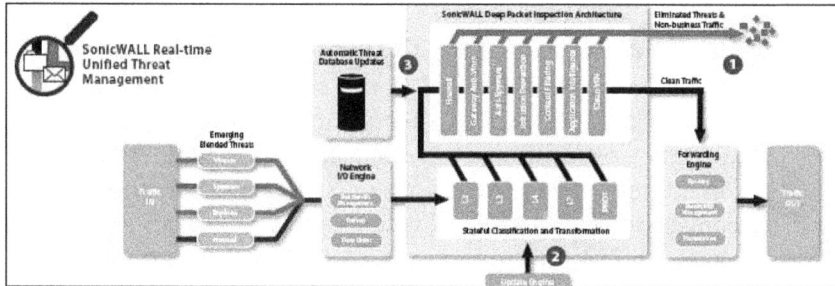

Figure 4. SonicWALL's UTM Architecture from 2007

Securing Ethernet Switches: Hardening equipment and establishing To ensure that all appliances have the same security configuration, a baseline security configuration is required The network infrastructure is safe. The technical settings will be the emphasis of this phase of the research. To obtain the best level of integrity, this should be done in Ethernet switches [8].

1. **Default options:** Network switch security is a critical technology, It's also one of the network's most important elements. The bulk of appliance factory settings are not changed, making them easy targets for burglars and hackers. In general, static protected device configurations In tiny transit networks, a protocol that is constantly changing that makes decisions based input is significantly more difficult to compromise [8].

2. **Attacks on the second layer:** Per-port MAC address security is enabled is the best practice for mitigating and preventing OSI layer 2 attacks. This capability is available on the majority of managed switches. Only the MAC addresses of approved hosts that have been bound to the network will be able to interact if this type of protection is enabled [8].

3. **Switch Management:** Using access lists, guaranteeing switch may be operated from certain nodes that the switch can be controlled from a number of different nodes, assuring that the switch may be controlled from a certain node,

ensuring that the switch can be, and There is no distinct text in the broadcast. For example, utilizing telnet is insecure since all data is sent in replaced by using the secure shell port instead of clear text, which can be readily intercepted, and completely restricts access to the device [11].

4. **Network Surveillance**: All of the above-mentioned security methods are monitors. But in some circumstances, additional safeguards are expected to ensure that consumers' access to the internet is secure. This is why the government and other security groups are concerned about it and have joined forces to In order to detect harmful software, keep an eye on the internet[16,17,18]. When something is discovered, they Endeavour to eliminate the threat as quickly as possible. as soon as possible in order to cause the least amount of damage feasible[19,20,21

6. Proposed Strategy for Protecting Small Business Network Infrastructure:

Our strategy to securing SME Network Infrastructure will be presented in this section that may be utilized to protect SME network infrastructure, Our recommended design for securing SME Network Infrastructure is shown in Figure 5. The proposed strategy emphasizes Management will be the primary line of defense (UTM), which is an abbreviation for zone-based inspection, To prevent local assaults, Basic preventative strategies (layer 2 security setup) should be considered. Rather than investing in separate firewall hardware, we recommend that SME's invest in Unified Threat Management. The company while staying within its budget. The majority of small businesses cannot afford to invest in the specialized device to protect them from a variety of threats. For VPN connectivity, use IPS/Gateway, Antivirus, Anti-Spam, Anti-Spyware, and IPSec will all be provided by UTMs. The appliance's initial setup, which will entail configuring all of the policies, as well as the technical setup, is the most difficult part of adopting UTM in a SME.

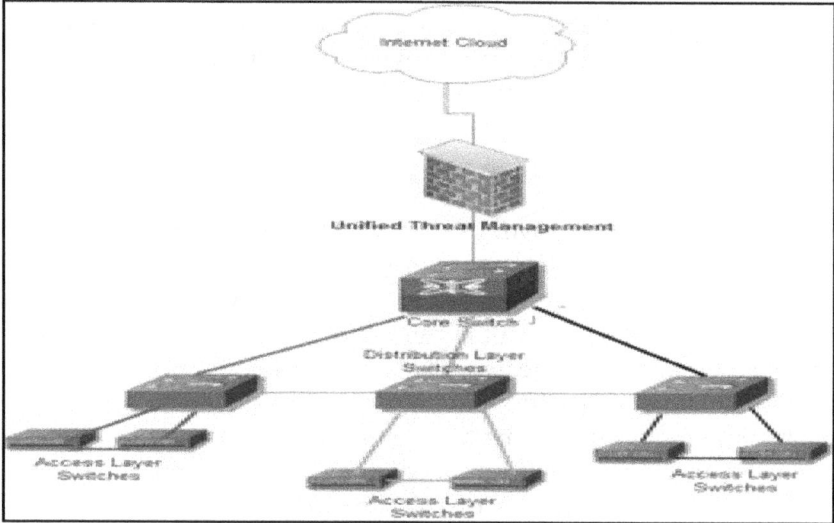

Figure 5. Diagram of the Network Proposed

Table 1. Comparison of specialized appliances against our proposed method

Comparison points of interest	Appliances with a Purpose	Our Methodology
Management Simplicity	Every appliance should be managed on its own.	a single point of contact for management
Configuration	Every appliance must be set up. via its interface.	Configuration in a single location
Appliances that connect security systems	It is necessary to establish mutual trust between Appliances	Single setup is not required.
Complexity of Configuration	Complex	GUI/Wizard based
Investment	In terms of hardware,	Gear can be replaced after a few years in the services offered (CGSS)
Defending against dangers	Only specific threats are considered threats.	Threats of various kinds

We put UTM gadgets in two (two) SME's in Oman as a test, and The infrastructure of the organization was safe, without any server or network infrastructure equipment failures

Conclusion

This study reviewed network security's significance, examined various types of network infrastructure threats, examined various approaches for mitigating network infrastructure vulnerabilities, and offered a way for safeguarding the network infrastructure of small and medium-sized businesses. That method implied that the initial level of protection for the Unified Threat Management can the network depends on the connections among every tier of the Defend your network with a distribution switch that allows for zone-based inspection and monitoring of the network against any potential threats. The UTM also comes with anti-spam, anti-virus, IPS/Gateway, anti-spyware, and IPSec for VPN communication. With their restricted Most small businesses can't afford a dedicated firewall because of their limited funds equipment to protect against several levels of attacks. The appliance's initial setup, which will entail configuring all of the technological setup as well as, all of the policies, is the most difficult part of adopting UTM in a SME. To guarantee maximum The company provides security against the most recent vulnerabilities and attacks and should constantly up-to-date signatures and appliance firmware for the services, just like any other security appliance. To summarise, but any businesses are Many others will fall short in a number of ways despite their best efforts, including failing to prioritise security in all business. There is no internal auditing method, there is no budgetary consideration for relevant issues, and there are no ongoing skills. In the future, we will need to fine-tune our recommended UTM approach for small and medium-sized businesses, continue to work on various frameworks for enterprise-level companies and put our strategy to the test in a high-stakes security setting.

References

[1] Li, Y., Huang, Gq., Wang, Cz. et al. Analysis framework of network security situational awareness and comparison of implementation methods. J Wireless Com Network **2019,** 205 (2019).

[2] B, Sachin., S, Yogesh kumar., J. Sahil, Modern Network Security, J. Contemporary Research In India, April 2021.

[3] T, Ciza., Computer Security Threats In Computer Security Threats, volume pp 1-20, September 2020.

[4] J. FitzGerald and A. Dennis, 2021 Business Data Communications and Networking, 14th Edition, Wiley Publisher.

[5] D. Moore., G. M. Voelker., and S. Savage, (2020) Inferring Internet denial-of-service activity," ACM Transactions on Computer Systems (TOCS), vol. 24, no. 2, pp. 115-139.

[6] Grimes, Roger [2017] Password Hacking, Wiley Online Library. https://doi.org/10.1002/9781119396260.ch21

[7] M. Hinson [2021], "Social engineering techniques, risks, and controls," vol. 37, pp 32- 46.

[8] R. Wanger [2001] "Securing Network Infrastructure and Switched Networks," SANS Institute InfoSec Reading Room, pp 1-20.

[9] I. Poynter; and B. Doctor, "Boyond The Firewall: The next level of network security," White Papers Series, January 2003. http://www.stillsecure.com/docs/StillSecure_BeyondtheFirewall.pdf, Date Visited: June 2022.

[10] I. Kenneth and F. Stephanie, "Network Firewalls [2004]," Enhancing Computer Security with Smart Technology, P. V. R. Vemuri, ed., CRC Press, University of California, 2004.

[11] D. X. Song, D. Wagner, and X. Tian, [2001] "Timing analysis of keystrokes and timing attacks on SSH," Proc. SSYM'01 10th conference on USENIX Security Symposium, 2001, pp. 1-25.

[12] I. Security, [2010] "The Complete Guide to Securing Your Small Business".

[13] R. Eubanks, "Application Firewalls: Don't Forget About Layer 7," SANS Institute InfoSec Reading Room, pp. 1-16. https://www.sans.org/white-papers/1632/

[14] Sonic WALL, "Unified Threat Management," Network Security, https://www.sonicguard.com/Solutions-UTM.asp

CHAPTER 4

MANET (Mobile Ad hoc Networks): An Overview, Applications, Challenges and Routing Protocols

Neetika Bansal and Kamaljit Kaur
Computer Science, Punjabi University TPD Malwa College, Rampura Phul, Punjab, India

1. Introduction

The advent and indispensable usage of MANET (Mobile Adhoc Network) has increased considerably in past decades. The structure of MANET is flexible. Moreover, these nodes are free to move randomly as the network topology changes frequently. Ad hoc wireless networks present various difficulties for mobility management. The topology of the network is constantly changing due to mobility. The routing protocols must dynamically readjust to such changes in order routes. This chapter also discusses the many challenges, applications and uses of MANETS.

The mobile hosts inside a network can communicate with one another whatsoever may be the infrastructure. In past decades, mobile ad hoc networks (MANETs) have become indispensable part of the communication networks. The fundamental benefit of a MANET is that the network may be instantly formed whenever and whenever. A MANET can therefore be employed in a combat area, a remote location, etc. MANET is a decentralized wireless network. It doesn't depend on a managed infrastructure access point or routers in wired networks are an example of existing infrastructure. Therefore, it is ad hoc. MANETs are wirelessly connected to numerous networks and are portable.

Due to the frequent changes in network topology, MANET nodes are free to migrate anywhere they like. The communication is passed to other designated nodes in the network and every node behaves like a router Due to emerging 3G and 4G activities, Mobile Ad Hoc Networks (MANETs) are gaining importance (Sesay, Yang & He, J, 2004). Many companies that require wireless roaming can benefit greatly from MANETs because of the wide range of applications they can be used for, including the battlefield, emergency services, and catastrophe finding. Intrusion detection systems or cryptography are frequently used to secure multi-hop ad hoc networks (Korba, Nafaa & Ghanemi, 2016).

By encrypting data and using node authentication, cryptographic techniques can defend ad hoc networks from outside attackers. However, these methods cannot prevent insider assaults, they are resource intensive, and they have additional issues like key management and issuance (Mirza and Bakshi, 2018).

Through network activity monitoring and analysis, intrusion detection and prevention systems can identify harmful actions taken by internal or foreign attackers. Each node has an IDS agent in a cooperative architecture that conclusive detections must be resolved. A type of cooperative design appropriate for multi-layered networks is hierarchical IDS. The network is organized into clusters in this architecture, and some nodes act as cluster heads. While cluster leaders carry out global detection, each cluster member does local detection (Sen, 2010).

Any collections of networks that are open to connections are often referred to as ad hoc networks. A group of movable nodes that instantly establish a network without a set topology is known as an ad hoc network. There are no base infrastructures employed in the conventional network in the instantaneously constructed ad hoc network. Despite this, it works with the conventional networks. In the world of computers, wireless networks have gained popularity. terminals, multi-hop routing, shared physical media, and autonomous terminal.

In MANETs a group of two or more wireless nodes that may communicate with one another without the assistance of a controlling node. Because the nodes in a MANET are movable by nature, the topology of the network frequently changes. An ad hoc network's performance, however, tends to suffer as its users move around. The traffic control overhead needed to keep accurate routing tables in the presence of mobility is one cause of this degradation. Different mobility patterns will have varying effects on how well various network protocols perform.

To attain the greatest performance in each situation, it is crucial to research how different network protocols are affected by mobility patterns. Flexible connections between clients in different locations are made possible by wireless networks. Without a connected connection between the customers, the network can also be expanded anywhere or within a structure. There are two types of wireless networks, namely: infrastructure networks and Ad Hoc networks.

1.1 Infrastructure Networks

This type of wireless infrastructure network depends on a third stationary party, but it also features an architecture that enables wireless stations to communicate with one another. As shown in Figure 1, A central coordinator between all nodes is an access point (AP). Through AP, any node can join the network. The base station is informed by the source node whenever it wishes to communicate with a Destination node. The basic overhead incurred includes routing table maintenance.

1.2 Infrastructure less Networks or Ad Hoc Networks

Infrastructure less Networks is also known as a Mobile Ad Hoc Network.

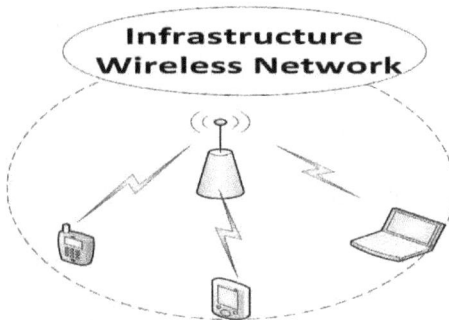

Fig 1: Infrastructure Networks

(MANET), and it allows users to communicate with one another without the help of a central administrator as shown in figure 2. Ad hoc networks normally do not have fixed topology and a centralized hub for collaboration. Consequently, the source and destination nodes communicate with one another by transmitting packets, which is a more advanced method of communication than Mobile Adhoc Network.

Each node acts as a relay when necessary to carry out specific tasks like routing and security. The nodes that make up a MANET should cooperate and interact with one another The packet must be resent through one or more intermediary nodes if a node wants to convey information to the node not falling in its communication domain.

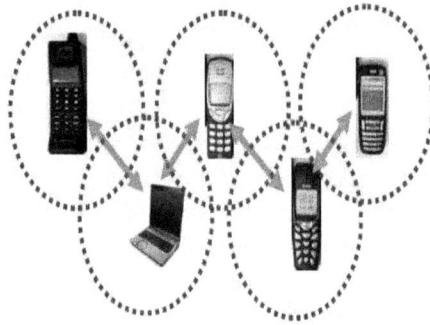

Fig 2: Mobile Ad Hoc Network

Every mobile node in a MANET is a separate node that can be used as a host and a router also. The network topology in a MANET undergoes some changes at unexpected periods. The nodes have complete freedom to travel at different speeds throughout the network domain. Because of this, the nodes in the MANET create their network as they move around and dynamically construct routing between one another. The MANET nodes are often portable, with limited CPU capabilities, low power storage, and small memory sizes, allowing users to carry them wherever they go. Any entity that has the right tools and sufficient resources can access the wireless communication medium. Access to the channel cannot, therefore, be restricted. Multihop routing is a kind of radio network communication technique used when the radio network coverage of a single node is not possible. Consequently, a node can use other nodes as relays to reach a certain destination.

A dynamic network topology, which sets the neighbor connections to be maintained by the network nodes, is the foundation of a peer-to-peer network. due to bandwidth restrictions and changeable capacity links, wireless communications have lower capacity than infrastructure networks.

2. Related work

With signature-based detection, system activity is compared to patterns or signatures of known attacks. It is trustworthy and has a low rate of false positives, but it is unable to identify new attacks. Specification-based detection identifies can identify unidentified assaults, but establishing the specification takes time. For the first time, the specification-based IDS for MANETs was put forth by Tseng et al. in 2003. To find run-time spec violations, they employ distributed network monitors which were supposed to monitor the nodes. Finite state machines (FSM) are used by network monitors to specify the proper AODV routing behavior. A request reply flow is kept track of for each Route Reply (RREP) and Route Request (RREQ) communication within the scope of the network monitor.

Panos et al. (2010) presented IDS incorporating a random walk-based architecture. It can very well supervise the network, transport and data link layers of the protocol stack. In addition, a multi-layer detection engine has also been suggested. To monitor node behavior and identify potential assaults that might occur in the visited node, a group of self-contained Random Walk Detectors (RWDs) moves randomly between nodes in the network. Attacks that violate specifications can be found by RWDs at many layers. However, when the number of RWD increases, the migration process for RWD results in substantial data transmission and additional communication overhead. Specifications for the suggested protocol seem to be lacking.

Panos et al. (2014) suggested the specification-based IDS SIDE track the actions of the host node. SIDE can find specifications that are broken, real time attacks and highly accurate detection. However, it is dependent on hardware support and makes use of Resource intensive encryption and authentication services.

Shakshuki et al. (2013), proposed Enhanced Adaptive ACKnowledgment (EAACK). A digital signature is used in the acknowledgement-based method EAACK to validate and authenticate the acknowledgement packets. It requires end to end acknowledgment for every packet sent. EAACK exhibits strong detection rates for a specific type of attack. Even the most well-known assaults like flood and black hole attacks coulnot be detected by it, and it generated a substantial amount of overhead (acknowledgement packets). Furthermore, even though the authors have brought up the issue of the additional resource use costs brought on by digital signatures, they provided no recommendations for how to reduce these costs.

The IDS responds to an attack after it has already happened; it is powerless to stop it. Tasks that require a lot of time, bandwidth, and resources include ongoing data collection, repetitive training, attack inference, and knowledge base administration. Workload, categorization accuracy, and energy usage should all be traded off. Additionally, creating and adding a rule for the new assaults increases the likelihood of producing misleading attack signatures. Intrusion detection and adaptive response (IDAR), is a distributed intrusion detection system based on signatures that Alattar et al. (2012) suggested (Clausen & Jacquet, 2003). In order to confirm incursion, IDAR gathers evidence from OLSR collected logs and launches a thorough cooperation investigation depending on the amount of suspicion around the activity. IDAR keeps comparing all the logs with predefined signatures to identify attack patterns.

Even while IDAR can only identify specific threats that are in its signature database. Additionally, IDAR is susceptible to rogue nodes that could transmit false information while conducting an inquiry (blackmail attack). Additionally, activities like gathering and analysing logs need a lot of resources (memory and bandwidth). Anomaly based IDS, suggested in (Mitrokotsa & Dimitrakakis, 2013), have been implemented using a variety of techniques, the majority of which are based on artificial intelligence techniques.

Jabbehdari et al. (2012) suggested an intrusion detection system to identify DoS assaults in MANETs. This system was based on neural networks with a combination of two techniques Negative Selection (NS) and Artificial Bee Colony (ABC) algorithms developed by Barani et al. (2012) as an anomaly-based IDS called Bee ID that can identify a variety of threats.

This chapter discusses the applications of MANET in Section 3, and then in Section 4 different Routing Protocols are discussed. It also explores some of challenges encountered while using MANETs in Section 5.

3. Applications of MANETS

The MANET is currently so widely used that it has numerous practical applications in a variety of industries. Mobile ad hoc communications are commonly used in society, such as taxis, boats, sports stadiums and small aero planes. Ad-hoc systems are self-regulatory and can connect to a moment and transitory mixed media (Gupta, Verma & Sambyal, 2018). The primary uses of MANET are as follows:

3.1 Emergency services

In many emergency operations, MANET is used. When calamities take place, MANET plays a crucial role in the rescue and identification of casualties by substituting the established infrastructure. It is especially helpful in cases of environmental catastrophes. It is also utilized in firefighting and law enforcement. It is also utilized during search and rescue efforts. It is also helpful in the medical domain as well.

3.2 Commercial and civilian environment

The business and private sectors use MANET. It is utilized where electronic payments can be made at any time and anywhere. It is helpful in business for offices and mobile devices to maintain and updating the databases. It plays a crucial role in automobile services by guiding drivers. Henceforth, avoiding accidents and transmit weather and road conditions to vehicles connected via networks. Wherever some protection is required, MANET helps out in timely conveyance.

3.3 Military services

Nowadays, computers are apparently used in military equipment by maintaining strong connectivity amongst the soldiers, vehicles and information headquarters while utilizing standard network technology. This field has given weightage to the fundamental principles of ad hoc networks. Through specially appointed technical admin, the military can efficiently maintain data regarding the personnel, vehicles and military headquarters. Hoebeke et al. (2004).

3.4 Online security

MANET is necessary for security in the residential areas. The devices used as such are intelligent sensors, security cameras and automatic door locking systems. In addition, data tracking of ambient variables, animal movements, and chemical and biological detection are accomplished through MANETs.

3.5 Home and enterprises networking

MANET is helpful in home/office wireless networking. In conference and meeting rooms, it is utilized for data exchange. Personal networks and personal area networks (PAN) are utilized with it (PN). When lot of construction is going on, big enterprises also get benefitted by MANET.

3.6 Entertainment

Multimedia, games are just one of the entertainment applications of MANET. that leverage MANET. It is utilized also in Wireless P2P networking, Robotics, security online, A/M, F/M etc.

3.7 Context aware services

Call forwarding and mobile workspace are two add-on services where MANET is helpful. Additionally, it is utilized in information services like time and location-based services. It is employed in the tourism industry for information exchange and geospatial data.

3.8 Education

With the advent of COVID, the necessity of online classes has contributed to a great hike with the use of MANETs. Multimedia games are one of the entertainment applications of MANET.

3.9 Business applications

Ubiquitous computing is a strong Manet's business application. Data networks are expandable from typical range of physical infrastructure to transmit data for others. Networks can be made more accessible and user-friendly. Short-extended MANETs encourage the intercommunication between different handheld devices like tablets, mobiles and wearable computers.

4. Routing protocols in MANET

Routing is the process of choosing routes via a network to move data packets from a source node to a destination node via a variety of nodes. The routing algorithm is a collection of rules that track the network's operations for making up a routing protocol. The major challenge encountered in MANETs is that the routing protocols must adapt to changes in the topology of the network. Network adhoc Proactive, reactive, and hybrid routing protocols are the three categories into which routing protocols can be separated.

A number of wireless mobile computers (or "nodes") that work together by helping each other forward packets to enable communication outside the direct wireless range is known as a mobile ad hoc network. Transmission Ad hoc networks don't need established network infrastructure like base stations or access points or centralised administration. An autonomous collection of mobile users known as a MANET can communicate via wireless networks that are only moderately fast. Because the nodes are movable, the network architecture may change quickly and unpredictable over time. Such a network might function independently, or it might be linked to the wider Internet. Certain characteristics of MANETs include dynamic network topology, limited physical security, energy constrained operation, links with changeable capacity and bandwidth, and frequent routing updates (Gupta and Verma, 2011).

4.1 Proactive Routing

Proactive routing protocols are mainly table-driven. Each node in proactive routing uses the tables to maintain the routing information. Any changes to the network topology are propagated across the system to ensure that the network remains proportional. They make an effort to keep the routing data for the entire network accurate and consistent. It lessens communication lag and enables nodes to identify which nodes are accessible or present in the network fast. These protocols constantly keep track of the routes that amongst different nodes.

By sharing topological data across network nodes, these protocols continuously learn the topology of the network. As a result, route information is always available when it is needed to get somewhere. Information about the routing status is tracked differently by different protocols. In order to keep current routing, information table-driven is frequently used to describe these protocols. These protocols constantly work to keep all communication mobile nodes connected, attempting to do so before a route is required.

The tables are synchronized by exchanging periodic route updates. Adhoc mobile networks use the table-driven, Bellman-Ford algorithm-based routing technique known as Destination-Sequenced Distance Vector Routing (DSDV). The Distance Vector for Destinations in Sequence (DSDV)The Bellman Ford Routing Algorithm is the foundation of the routing protocol, with certain modifications to make it loop free, for example. Each device in this network keeps a routing table with entries for every other device. Each device occasionally broadcasts routing messages to keep the routing table fully updated. A neighbor device compares the current connection cost to the device with the corresponding value stored in the database when it receives the broadcasted routing message and is aware of it. If modifications were discovered, the values are updated and the distance of the route are recalculated. A proactive uni cast routing mechanism for MANETs is the Wireless Routing Protocol (WRP). The Bellman Ford algorithm is used by WRP to determine pathways. The protocol offers reliable message exchanges. The Optimized Link State Routing System (OLSR) is a proactive link state routing protocol that figures out the link status information and then spreads it across the mobile adhoc network. Using the shortest hop forwarding paths, every single node computes the next hop destinations for every other node in the network.

4.2 Reactive Protocols

Dynamic Source Routing(DSR) and Ad-hoc On Demand Distance Vector Routing (AODV) are two examples of reactive routing technologies.

Reactive routing systems create routes as needed, rather than maintaining them. By saturating the network with Route Request packets, a reactive protocol locates a route on demand. The benefits of these methods include: Maintenance of the global routing table is far less expensive than with proactive protocols. Prompt response to node failure and network reorganization. Reactive protocols still have the following primary drawbacks, despite becoming the standard for MANET routing: Long route finding delay times Excessive flooding may cause a network to become congested. For MANET, there are numerous reactive routing protocols. In this section, we only present one novel protocol (ODCR) and three popular ones (DSR, AODV and DYMO). (Perkins Royer and Das, 2003) developed it together at the Nokia Research Center, University of California, Santa Barbara, and University of Cincinnati.

Both unicast and multicast routing are possible with AODV. Routing Sources Dynamically (DSR) is a routing protocol for wireless mesh networks. In that respect, it resembles AODV. For lengthy paths or large addresses, like IPv6, this could lead to a substantial overhead. DSR defines an optional flow id option that enables hop by hop packet forwarding to avoid using source routing. OnDemand Dynamic MANET Routing (DYMO) have been proposed as the DYMO routing protocol as an improvement to the current AODV protocol (Perkins & Chakeres, 2005).

It also describes itself as the AODV or ADOVv2's successor and continues to be updated today. While DYMO does not offer any new functionality to the functionality of its predecessor, AODV, operation is also far more straightforward. Routes are calculated in DYMO, a reactive protocol, as and when they are needed. DYMO, not AODV, Route maintenance and discovery are the fundamental processes. To reach a destination for which it lacks a viable path, route discovery is carried out at the source node. Additionally, route maintenance is carried out to prevent current routes from being deleted from the routing database and to lessen packet loss in the event of a route break or node failure (Palaniammal and Lalli, 2014).

4.3 Hybrid Protocols

Both the proactive and reactive procedures' positive aspects are combined in hybrid protocols. In the network topology, it combines groups nodes into zones to maintain routing information. For routing packets across several zones, a reactive technique is used. One of the varieties of hybrid routing protocols is Zone Routing Protocol (ZRP). For packet switched networks, the routing method Ant Net was first presented. A random receiving end is chosen for sending ant packets to the receiver node. The cheapest communication path between the sending and receiving nodes is determined by ant agent.

Each member node in the network stores two different sorts of tables; the first is a routing table and the other contains data on how traffic is distributed throughout the network. A backward ant packet (BacAnt) follows the same path that the forward ant packet (ForAnt) took to record any changes in the network's state. The forward ant packet (ForAnt) goes through the network's communication paths in search of the best route. With the node id, ForAnt records the time it took to get to a node in its buffer. The next hop to traverse is chosen based on computed probabilities. Once at the destination node, ForAnt changes into BacAnt, which uses the stack that ForAnt had saved to travel back. Information is also updated at each node's routing table with probability of the state of the network as the BacAnt moves across it. The packet is deleted when BacAnt reaches the sender node so that data can be transferred via the predicted path. To address delay and congestion, BacAnt has a higher priority in the AntNet network than data packets and ForAnt.

In the ant colony-based routing algorithm (ARA), ant packets are only sent when necessary and are then broadcast throughout the network using the same method as the ad hoc on demand distance vector (AODV). A hybrid algorithm called AntHocNet is supported by ant colony optimization (ACO) methods (Sharma and Ram Kumar, 2017). Only those target nodes must be taken into consideration for the current session which are included in this algorithm's route knowledge union. Reactive and proactive ant packets are the two types used by AntHocNet. To determine the best path for communication, reactive ForAnts and BacAnts are sent throughout the network. Reactive ant packets are sent to establish the route since the sender node lacks route information when the transmission session begins. Based on the distance travelled and the number of hops, certain paths are disregarded. Ant packets are typically unicasted, and they employ the same method as reactive For Ants to select the next node based on pheromone concentration.

A hybrid ant colony optimization routing algorithm for mobile ad hoc networks (HOPNET) technique combines reactive transmission across communities. The HOPNET network is separated into zones, with each zone's size determined by the radius's length and measured in hops. A node may or may not be a member of more than one network zone. In order to identify zones, hello packets are sent throughout the network. ZRP is made up of two sub-protocols: Inter zone Routing Protocol (IERP) and Intra Zone Routing Protocol (IARP).

A mesh network of computers communicating by digital radio uses the Order One MANET Routing Protocol (OORP) to locate one another and transfer messages along a reasonably efficient path. While most other protocols can only manage a few hundred nodes, OORP can handle hundreds. To reduce the overall number of transmissions required for routing, OORP employs hierarchical algorithms. Link based Routing Protocol of preference (PLBR) Protocols for reactive routing Fundamental idea Each node keeps two tables updated: NNT and NT A subset known as Preferred List (PL) K is chosen by each node: the PL's dimensions Optimal List creation Neighbor based Degree.

The Preferred Link Algorithm separates its neighbor nodes into reachable and unreachable based on the degree of their proximity.

The weight assigned to a node by the weight based preferred link method is based on the temporal and spatial stability of its neighbors.

5. Challenges encountered

Despite the qualities MANET possesses, there are still certain difficulties that must be overcome. Over the past few years, MANETs have emerged as a major area of study these days. In spite of numerous applications that MANETs provides us, there are still some challenges that have to overcome (Kumar and Mishra, 2012).

5.1 Limited bandwidth

Wireless link has much lower capacity than infrastructure networks is one of the problems with the MANET. In addition, the actual throughput of wireless communication is quite lower than the maximum transmission rate of a radio after several access effects like loss, noise, and interference situations, etc. are considered.

5.2 Dynamic topology

Another issue with the MANET is that the wireless connections have lower capacity than infrastructure networks, accounting to various access effects like loss, noise, interference conditions, etc., the real throughput of wireless communication is typically significantly lower than the maximum transmission rate of audio. The confidence gets diminished if nodes are discovered to be compromising. MANET nodes regularly fuse with or dispense the network, move independently, and depart from the network at their inclination. Since nodes can move throughout the network easily and seek security at any time and makes integrating security very challenging (Ali and Kulkarni, 2015).

5.3 Routing Overhead

Nodes frequently reposition themselves within wireless ad hoc networks. As a result, the routing table contains some undesired routes, which incurs huge routing overheads. Since the nodes disperse continuously, table driven routing protocol i.e. the reactive routing protocol can be used. Again Multicast routing becomes challenging as the nodes move freely and multi-cast tree no more remains static.

5.4 Packet loss due to transmission errors

Due to issues including increased collisions, unidirectional links, hidden terminals, frequent path interruptions and interference due to node mobility. Hence, ad hoc wireless networks face substantially higher packet loss.

5.5 Mobility-induced route changes

Because nodes move around in ad hoc wireless networks, the network architecture is quite dynamic. Consequently, a continuous Session experiences numerous path breaks. This circumstance frequently causes repeated route adjustments.

5.6 Battery constraints

The limitation of the battery reduces the efficiency of MANET. To retain the portability, size and weight of the device, power source restrictions are faced by the devices utilized in the networks. To retain the portability, size, and weight of the device, devices utilized in these networks are subject to power source restrictions.

5.7 Security threats

There are specific security issues with MANET. Short wireless transmission range makes the design of the network face additional security difficulties. The wireless channel is leaky and ad hoc network functionality is built through node collaboration. Thus, making MANETs more challenging.

5.8 Quality of Service (QoS)

Because of the environment's mobility, the quality of service is regarded as one of MANET's greatest challenges. It is challenging as the communications quality in a MANET is inherently unpredictable. To enable the multimedia services, an adaptive QoS must be provided in addition to the traditional resource reservation (Goyal, Parmar and Rishi, 2011).

5.9 Scalability

If a network can deliver an adequate quality of service to numerous nodes, it is said to be scalable. To arrange distant Quality of Service (QOS) levels for devices in an environment is very challenging. Due to stochastic nature of MANET stable QOS cannot be guaranteed. There is a straight need for adaptive QOS to be implemented (Bang and Ramteke, 2013).

5.10 Network configuration

The entire MANET infrastructure is not stable, and it causes the variable links to connect and disengage dynamically. Sometimes MANETs cannot n interact via co-existing routing protocols.

6. Conclusions and future scope

Adhoc networks have a very promising future since they promise affordable, anytime communications. There is still a significant amount of study and implementation work to be done before those envisioned possibilities become reality. Mesh architecture and big size are currently the general trends in MANET. It is necessary to increase bandwidth and capacity, which suggests that a higher frequency and improved spatial spectral reuse are also necessary. Another difficult problem shortly that can already be predicted is largescale ad hoc networks.

In past, the mobile communication has made some significant advancement. As a result, there are now many opportunities available in the field of ad hoc networks. A temporary network is created by a set of wireless mobile hosts called a MANET without the need for centralised administration or backbone support services. Less effort is required thanks to the self-healing mechanism. MANET must reorient itself to encircle any node that steps outside of its coverage area.

MANET has proven to be a flexible network in the modern day, however, it is quite unreliable since it is vulnerable to attacks, or is less resistant to attacks. To enhance the study in this area, the purpose of this paper is to comprehend the difficulties and applications of MANET. Our understanding of the study has led us to believe that

Mobile Ad-hoc Networks (MANETs) will be a crucial piece of infrastructure for the development of a future omnipresent society. Since it is extremely difficult to construct extensive and realistic test beds in the real world for performance evaluation, designing MANET protocols and applications is a highly challenging undertaking.

References

Alattar, M., Sailhan, F. & Bourgeois, J. (2012). Log-based intrusion detection for MANET. In Wireless Communications and Mobile Computing Conference (IWCMC), Limassol (pp. 697- 702). USA: IEEE.

Ali, A. K. S., & Kulkarni, U. (2015). Characteristics, applications and challenges in mobile Ad-Hoc networks (MANET): overview. *Wireless Networks*, *3*(12), 6-12.

Bang, A. O., & Ramteke, P. L. (2013). MANET: History, challenges and applications. *International Journal of Application or Innovation in Engineering & Management (IJAIEM)*, *2*(9), 249-251.

Barani, F., & Abadi, M.I. (2012). BeeID: intrusion detection inAODV-based MANETs using artificial bee colony and negative selection algorithms. The ISC International Journal of Information Security, 4(1), 25–39.

C. E Perkins, E. M. Royer, and S. Das, "Ad hoc On-demand Distance Vector (AODV)," RFC 3561, July 2003

Chakeres, C. Perkins, Dynamic MANET On-demand (DYMO) Routing, RFC draft, Boeing, Nokia, February 2008.

Clausen, T., Jacquet, P. Optimized link stat e routing protocol (OLSR). 3626. IETF RFC, October 2003.

Goyal, P., Parmar, V., & Rishi, R. (2011). Manet: vulnerabilities, challenges, attacks, application. *IJCEM International Journal of Computational Engineering & Management*, *11*(2011), 32-37.

Gupta, A. K., Sadawarti, H., & Verma, A. K. (2011). Review of various routing protocols for MANETs. *International Journal of Information and Electronics Engineering*, *1*(3), 251.

Gupta, A., Verma, P., & Sambyal, R. S. (2018). An overview of MANET: features, challenges and applications. *International Journal of Scientific Research in Computer Science, Engineering and Information Technology*, *4*(1), 122-126.

Hoebeke, J., Moerman, I., Dhoedt, B., & Demeester, P. (2004). An overview of mobile ad hoc networks: applications and challenges. *Journal-Communications Network, 3*(3), 60-66.

Jabbehdari, S., Talari, S.H. & Modiri, N. (2012). A Neural Network Scheme for Anomaly Based Intrusion Detection Systems in Mobile Ad Hoc Networks. Journal of Computing, 4(2), 61-66.

Korba, A. A., Nafaa, M., & Ghanemi, S. (2016). An efficient intrusion detection and prevention framework for ad hoc networks. *Information & Computer Security.*

Kumar, M., & Mishra, R. (2012). An overview of MANET: history, challenges and applications. *Indian Journal of Computer Science and Engineering (IJCSE), 3*(1), 121-125.

Mirza, S., & Bakshi, S. Z. (2018). Introduction to MANET. *International research journal of engineering and technology*, 5(1), 17-20.

Mitrokotsa, A. & Dimitrakakis, C. (2013). Intrusion detection in MANET using classification algorithms: The effects of cost and model selection. Ad Hoc Networks, 11(1), 226-237.

Palaniammal, M., & Lalli, M. (2014). Comparative study of routing Protocols for MANETs. *International Journal of Computer Science and Mobile Applications, 2*(2), 118-127.

Panos, C., Xenakis, C., & Stavrakakis, I. (2010). A novel Intrusion Detection System for MANETs. In SECRYPT (Ed), International Conference on Security and Cryptography (SECRYPT), Athens (1-10). USA: IEEE.

Panos,C., Xenakis,C., Kotzias,P., Stavrakakis,I. (2014). A specification-based intrusion detection engine for infrastructure-less networks, Computer Communications, 54, 67-83.

Sen, S. (2010). Evolutionary computation techniques for intrusion detection in mobile ad hoc networks. Doctoral dissertation, University of York.

Sesay, S., Yang, Z., & He, J. (2004). A survey on mobile ad hoc wireless network. *Information Technology Journal, 3*(2), 168-175.

Shakshuki, E.M., Nan Kang, Sheltami, T.R. (2013). EAACK—A Secure Intrusion-Detection System for MANETs. Industrial Electronics, IEEE Transactions , 60(3), 1089-1098.

Sharma, I., & Ramkumar, K. R. (2017). A survey on ACO based multi-path routing algorithms for ad hoc networks. *International Journal of Pervasive Computing and Communications, 13*(4), 370-385.

68

CHAPTER 5

Cloud Computing and its Security Concerns

B Ravinarayanan, Dept. of Computer Science & Engineering,
MITE, Moodabidri, India
H R Nagesh, Dept. of Information Science & Engineering, AJIET,
Mangalore, India

Abstract

Computing resources are provided in demand by cloud computing, which helps many organizations to develop their business without investing in infrastructure. Infrastructures setup and its sharing are done in different ways among the service providers and its users. In the cloud computing architecture, the services are provided mainly by using the two layers, namely the front end and back end of its architecture with well-defined functionalities. Virtualization technology is the backbone of cloud computing. Cloud computing is an internet-based service and it is susceptible to various kinds of threats. It is important to understand the cloud architecture, various types of services proved by cloud computing and security concerns; while, these are important for effective adoption of the technology in the various business organization.

1. Introduction

Cloud computing is an internet-based service, it describes the delivery of several types of services like data and computing resources to its customers on-demand pay as you go technology using internet. The organizations can avail the cloud services as per their requirement. Shared resources can be provisioned and rereleased to the customers as their demand [1]. The Internet of Things (IoTs), Mobile Computing, Smart IT, and other IT applications have benefited from cloud computing. The key empowered the cloud computing is network virtualization technology. Independent multiple virtual networks are run over the common infrastructure using vitalization. Cloud computing eliminates the requirement of capital investment to start and run the organization. One important feature of the cloud computing, it provides the scalability to avail the right amount of service as per the requirement, this feature is called elasticity of services. As the technology develops the Cloud services upgraded to the new technology, since its services are offered worldwide, leading to provide better performance to its users. The organizations which use the cloud services concentrate more on the business than the IT infrastructure management leading to increased productivity. The mirroring of the data at multiple sites by the cloud providers network enhances the reliability. The policies, technology and control strengthen security against the threats in cloud environment.

2 Cloud computing characteristics

On-demand self-service: In contrast to manually fulfil the request of the customer by the administrator, in cloud customer can get the service offerings themselves. This is called as on-demand self-service. The cloud user request for the service to cloud service provider (CSP) to avail the service. This service request and fulfilling the requested service is automated in cloud services to quickly get an admittance the services. This on-demand self-service reduces the administrative burden on the CSP. This makes work of both the CSP and the service user more flexible and easier.

Broad network access: Because the service provided by the cloud are accessible through the network and it uses common protocols for accessing, allows diverse devices which runs the application locally or remotely such as laptops, personal digital assistants and mobile phones to access them. It stretches a provision of setting up a virtual office and connecting to the organization from anywhere around the clock.

Resource pooling: In a multi-tenant arrangement, the Cloud Service Providers (CSPs) share physical and virtual resources to service numerous customers. According to the customer's desire, the resources that have been pooled are assigned and reassigned.

Rapid elasticity: The cloud services often appear to be unlimited to the customers. The resources can be got rapidly when usage is more and released when usage is less, these provisioning is sometimes automatically also. This characteristic makes the cloud services for its widespread use.

Measured Service: The cloud service usage is a monitored controlled service with reporting, which makes the services rendered more transparent to the consumer and the service provider. The cloud uses automatic control and optimization of resources to its maximum utilization by a metering ability suitable to the type of service.

Scalability: The cloud's success and the reason for its extensive adoption is from its ability to handle one problem in a better way than any other technology, which is scalability. For the growth of any organization, it should be able to scale the infrastructure to the required extent to expand its operation fast, the converse of this may hamper its growth. This issue is significantly addressed by designing elasticity in cloud computing systems' services. Cloud computing provides flexibility for managing the services and computing resources on-demand with the cloud, and scale infrastructure as needed. At the same time, when necessary, this infrastructure can be limited without affecting investment and with greatest flexibility.

Performance: Cloud computing services also allow the hosting of the platforms, software, and database remotely to free up memory and computing power on the organization's machines. Further, cloud computing services are on-demand, you may have enormous amounts of computing resources available in a short time, with tremendous convenience and flexibility, and without making large expenditures. The most perceptible result is a substantial performance improvement.

3. Cloud Computing Models

The cloud infrastructure can be modelled in different ways; depending on terms of its ownership, scale, its purpose, nature of access and location of the server. Cloud services can be a private, public, community, or hybrid cloud configuration.

Private Cloud: The organizations set up the computational infrastructure in their premises and provides computational services. These computational services are shared among selected group of users using the private internal network or Internet rather than offering the services to general public, such cloud set-up is known as private cloud. Since it is set up and maintained by an organization and its services which are limited to a group of users related to that organization, can be termed a corporate cloud or internal cloud. Even though the users can avail many of the advantages of a public Cloud, such as scalability, elasticity, and self-service. It also provides customization and additional control over dedicated resources on the computing infrastructure, which are on premises. Furthermore, private clouds stretch better level of privacy and security, assuring that sensitive data and operations are not open to third-party, through the organizational firewalls and internal hosting. The private Clouds infrastructure is set up and maintained by the organization, and the IT department is accountable for the cost and operation of the private Cloud, due to these expenses of the private cloud in terms of management, maintenance, and staffing is same as that of traditional data centre ownership.

Two Cloud service models are offered by the private Cloud. The first one is Infrastructure as a Service (IaaS), wherein a business can rent infrastructure resources including computation, storage, and network. Another is Platform as a Service (PaaS), which enables businesses to deploy whole thing from simple cloud-based apps to futuristic corporate apps. A private cloud is shown in Fig. 1.1.

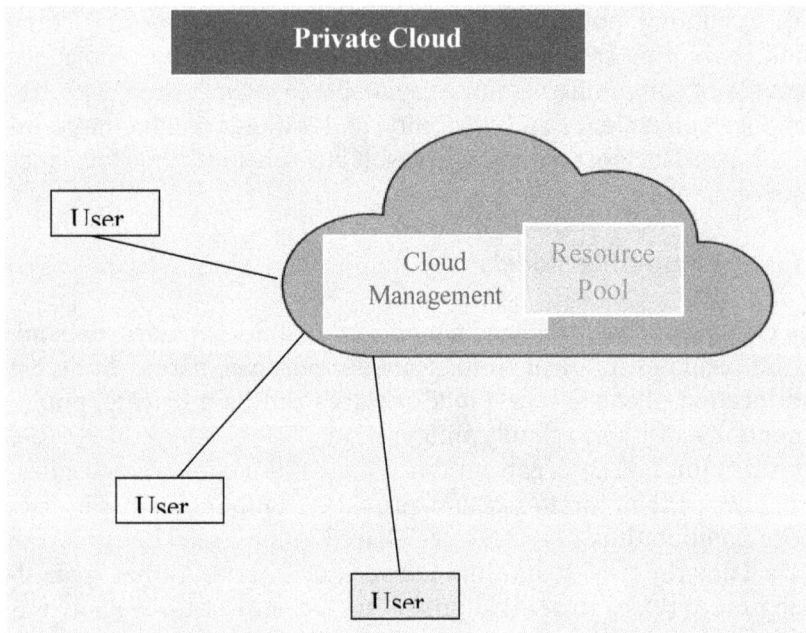

Fig. 1.1 Private Cloud

Community Cloud: A distributed system that brings together the services of different clouds to satisfy the needs of a specific community, industry, or business sector, such cloud models are referred as community cloud. NIST defines a community Cloud infrastructure as shared by community which works for common apprehensions (e.g., security requirements, mission, compliance considerations, and policy). In these case handling of the infrastructure is by the organisations or by a third party, and it can be off-site or on-site. Fig. 1.2 shows a community Cloud.

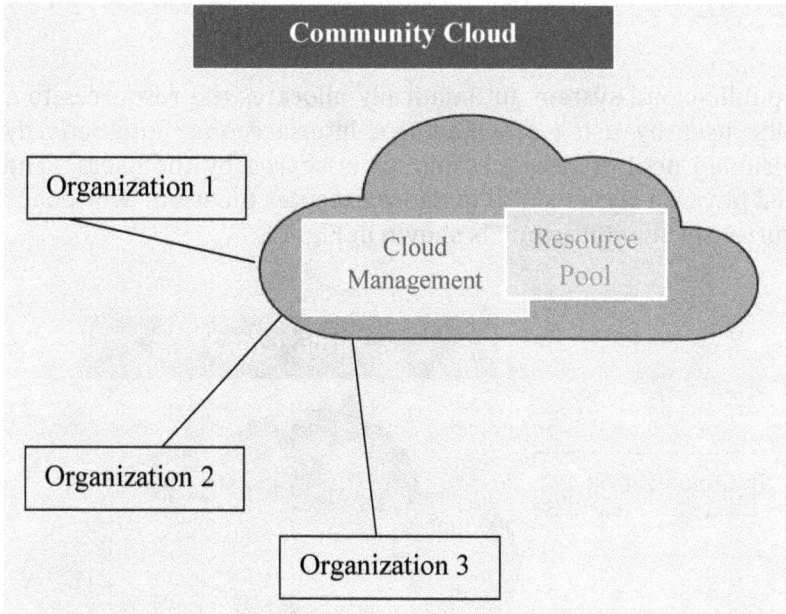

Fig. 1.2 Community Cloud

Public Cloud: One of the most significant revolutions in enterprise computing history is the rise and use of public cloud services. A cloud computing model wherein customers may access computing resources through the public Internet from a third-party service provider may be referred as public cloud. The services include ready-to-use software applications as the requirements of the customer, individual Virtual Machines (VMs), and whole enterprise-grade infrastructures and development platforms are the wide range of resources available in public cloud. Depending on the policies set by the cloud service provider, resources are made available for free, or the services are charged to its customer on a pay-per-use basis, or subscription.

The data centres operate the users workloads, which are owned and managed by the public cloud provider. For quick accessibility of applications and data are guaranteed by the service providers with necessary bandwidth and also, and they maintain the infrastructure and hardware. The underlying virtualization software required for providing the cloud service is also accomplished by the cloud provider.

The public cloud system automatically allocates the resources to a specific user by using a self-service interface, this automatically provisioned pool of virtual resources is shared by the users. The shared physical server's CPU instance executes the users' workloads concurrently. A public cloud is shown in Fig. 1.3.

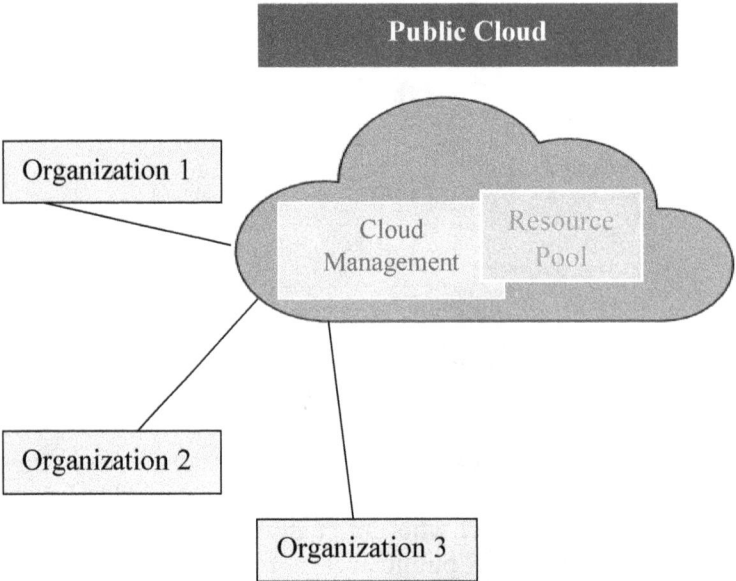

Fig. 1.3 Public Cloud

Hybrid Cloud: A single, flexible, cost-effective IT infrastructure that blends and unites private cloud and public cloud services offered by several cloud vendors that are used to provide cloud services to customers. This cloud strategy is known as a hybrid cloud. In hybrid clouds, the advantages of both public and private clouds can be blended. The organizations which use the hybrid cloud can share the data and apps in private and public cloud environments. When an organization's on-premise infrastructure needs more scalability, the public cloud's scalability can be employed to meet changing business requirements. Organizations can keep their sensitive data on their private cloud when using the public cloud.

4. Service Models of Cloud Computing

The type of right service requirements is decided by the user of the cloud computing services. Accordingly, cloud service providers offer mainly three types of cloud service models [2]: Software-as-a-Service (SaaS), Platform-as-a-Service (PaaS), and Infrastructure-as-a-Service (IaaS).

Software-as-a-Service (SaaS): SaaS enables the user to use a complete software product run and maintained by the service provider on a pay-per-use basis [3]. Software deployed as a hosted service and accessed via the Internet is a simple definition of SaaS[4]. This service offered by cloud provide scalability at the same time it also transfer considerable responsibilities from users to service providers, results in a variety of options to increase the efficiency and also increase in performance in some circumstance. The average SaaS user need not have awareness or control over the infrastructure used for the service [5].

Platform-as-a-Service (PaaS): Cloud computing service of this type delivers a development environment for application and its deployment. A customer may utilise the platform to create and manage his application and offer this to consumers over the Internet [6]. The customer has control over the apps that operate in the environment, without having control over the hardware, operating system, or network infrastructure. Thus, an application framework is offered as platform.

Infrastructure-as-a-Service (IaaS): IaaS provides an outsourced platform virtualization service. As a service, the user may regulate the environment. The user need not buy data centre space, network equipment, servers or software, the customers buy them as a fully operational installed application, storage, operating system, and perhaps load balancers and firewalls like network components. All these are provided as a service without setting up the infrastructure behind them.

5. Architecture of Cloud Computing

The architecture of cloud computing is designed using different components and subcomponents. All these components that join together to form a whole as a cloud. In cloud computing, security architecture is divided into two categories: frontend and backend, as well as a consolidation of service-oriented design and event-driven architecture.

The cloud computing environment is used by the interfaces and applications given in the frontend architecture; front end is visible to the user. The front end needs the communication with back end, this communication happens with the aid of the network like Internet. The middleware also allows the frontend to submit queries to the backend.

The service providers use the computing resources (generally termed as host) to provide services to its users, this cloud computing resources are termed as backend. The backend may be referred as the cloud itself. The backend also includes middleware, which allows devices to connect and communicate each other. The replies to inquiries from the front end are handled by the backend on time and secures the data from the front end. In addition to the security, this component of the cloud architecture also handles traffic management. In order to describe the brief about the cloud, the cloud computing architecture diagram is given in Fig. 1.4.

The cloud architecture is designed using various interlinked components and subcomponents which helps to run the activities of the cloud. The various components and its subcomponents are described in following subsections.

5.1 Cloud Computing Architecture Frontend

The interaction with cloud environments by the users is fully depicted in Frontend of Cloud Computing architecture. The front end is constituted with various subcomponents, which forms the user interface. The following are the main frontend components:
User Interface: This represents the activities that an end-user does to access a platform. Gmail and Google Sheets are some of the examples of user interfaces. The cloud may not cause any inconvenience

to the user to the service provided due to space or other limits, it takes the whole load. This is one of the significant advantages of cloud computing.

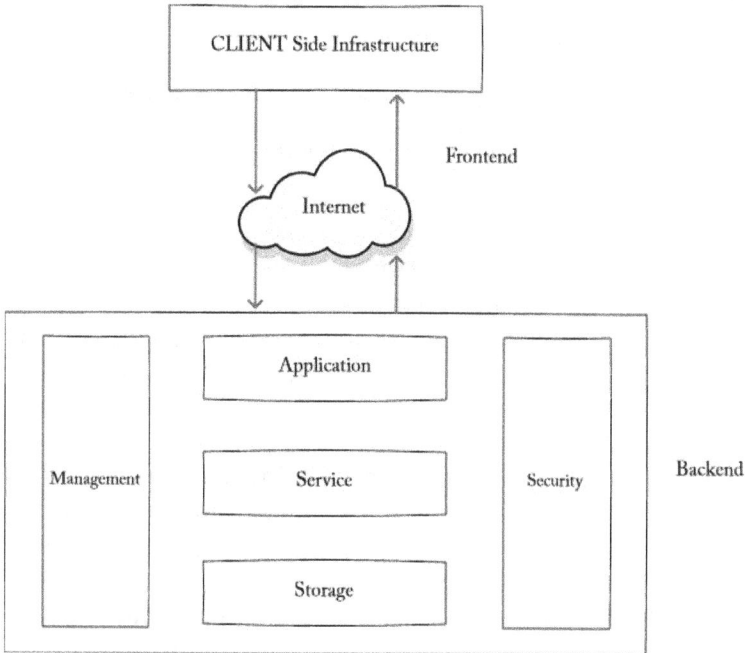

Fig. 1.4 Cloud Architecture

Client Infrastructure: The user interacts with cloud using a graphical user interface (GUI) in the client infrastructure. This is another front-end component of the cloud.

Software: The client-side applications and browsers, which is one frontend component represent the cloud software architecture.

Network: Client devices are connected to the cloud services to communicate frontend and backend of the architecture. The communication between the front end and back end of the cloud is established with the aid of a network channel. The network channel used here is Internet which communicate between front end and back end. The network might also be made available as a service, enabling clients to select their protocol and network route.

5.2 Backend of Cloud Computing Architecture

All functionality given to the frontend is supported by the backend. The backend is equipped with required hardware and the storage to provide the functionalities. It is the concern of the cloud service provider to create cloud computing backend. The backend of the cloud should provide an uninterrupted service, so the cloud infrastructure backend must be highly reliable. The backend architecture components are discussed in next subsections:

Application: Applications are the programmes or platform that the user wants to use in the cloud computing backend. As per the user's specifications, the backend recourses are synchronised.

Virtualization and Cloud Computing: Virtualization is the technique where we use single physical hardware resource to build many simulated environment or dedicated resources. The Information Technology (IT) environments that pool, abstract, and share the scalable resources are called clouds. The process of executing the workloads on a cloud is called cloud computing. Virtualization is technology used to set up a cloud computing environment. Virtualization expands the flexibility and usefulness of the hardware. With the aid of virtualization many programs and operating systems may work at the same time on a single server; as a result, virtualization expands the flexibility and usefulness of the hardware. It reduces hardware usage, prices, and energy consumption while allowing several operating systems and applications to run concurrently on a single server. Virtualization enhances the flexibility, effectiveness, and use of the primary hardware in physical server.Primarily, the virtualization manages the computer workloads more effective, efficient, cost-effective and scalable manner. This strategy helps in hiding the complexity of systems software and hardware [7-10].

Hypervisortechnology is often used to implement virtualization [11][12], the firmware or software element that virtualizes system resources. It allows to boost productivity, avoid data loss,give secure remote access and manage resources. Hypervisors are also known as Virtual Machine Monitor (VMM) [13]. The hypervisor is placed in between the guest operating system and hardware. A hypervisor helps the administrators to separate applications and operating systems from the underlying hardware. In Cloud computing, it is widely used since it permits several guest operating systems to function concurrently on a single host machine. Each guest operating run-on a single host machine is known as Virtual Machines. The resources such as RAM, CPU etc. are efficiently used by dividing between multiple VMs.Since guest operating system working is limited within VM, it unable to manage tasks in a cloud computing environment. The VMs in cloud are independent each other, if any one VM crashes that will not affect the other. This helps the migration of VM from one server to another without stopping them for load balancing and also in order to mitigate the DDoS attacks.

Hypervisor may be of two types based on which level it is implemented:
(i)   Type 1 hypervisor /native hypervisor/bare metal
(ii)   Type 2 hypervisor/hosted hypervisor
If the hypervisor is directly sits between the hardware and the guest operating system it is termed a Type 1 hypervisors. It is also known as bare metal as well as native hypervisor. Type 1 hypervisors directly operate on the physical machine by eliminating the need of a host operating system as indicated in Fig. 1.5.

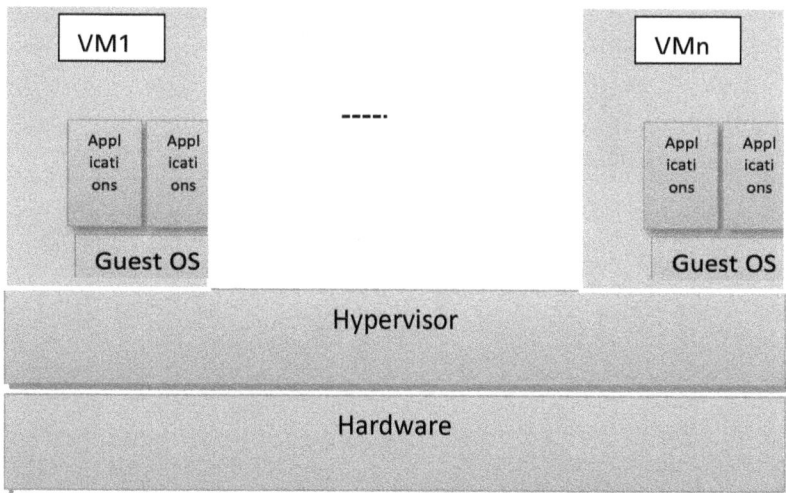

Fig. 1.5 Type-1 Virtualization

Citrix XenServer, Microsoft Windows Server 2012 Hyper-V VMware vSphere/ESXi, open-source Kernel-based Virtual Machine (KVM) and RedHat Enterprise Virtualization (RHEV) are examples of Type 1 hypervisors.

The Type 2 hypervisor sits on top of host operating system and controls virtual machines as well as supporting other guest operating systems, as indicated in Fig. 1.6. Hypervisor is layer between the actual computer hardware and the VMs. The additional layer is in the form of host operating system for Type 2 hypervisors. The examples for Type 2 hypervisors include VirtualBox, Microsoft Virtual PC, Oracle Virtual Box, and VMware Workstation. Host or guest machines describes various responsibilities in the hypervisor and the guest operating system controls each VM with its logical domain with a secure and isolated environment.

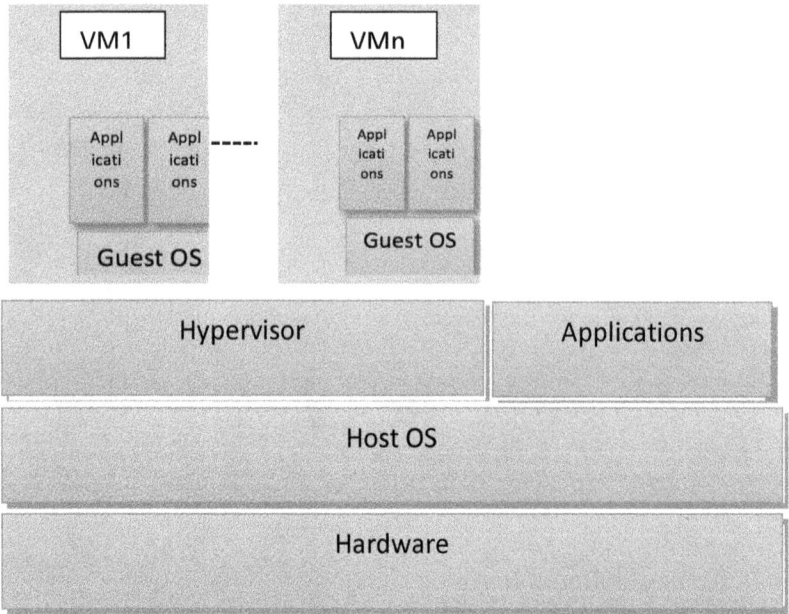

Fig. 1.6 Type 2 Virtualization

6. Cloud Computing Security

In cloud computing, the hypervisor oversees and supervises the whole virtualization process. The hypervisor is a piece of software or an operating system that acts as a traffic cop, ensuring that everything happens in the right sequence. The entire cloud computing infrastructure operates on a give-and-take basis, with an attacker gaining access as a result. But unlike a typical operating system, which is flexible in its dealings, a traditional operating system controls all hardware components, utilities and application software. However, no sound system exists, and hypervisor(s) that makes a sudden change to a cloud computing component would make it difficult for attackers to manage or obtain control of the entire virtual machine.

6.1 Security concerns for Cloud Computing

Majority of the organizations has fused cloud computing into their operations to varying degrees. The cloud is susceptible to various kinds of security threat. The wide adoption of the cloud leads security mechanisms, which is proficient of protecting cloud from top cloud security risks. The major security threats to cloud are as follows.

Misconfiguration: Misconfigured cloud security settings are the most common cause of cloud data breaches. Many companies' cloud security posture management systems aren't up to the task of protecting their cloud-based infrastructure. Many reasons which make the security insufficient. One such reason is designing of cloud infrastructure in such a way that its easy usability and easy data sharing, leads difficulty for the organisations to guarantee that data is only available to authorised users. Additionally, organizations that practice cloud-based infrastructure have, nonexistence of complete access and control over their infrastructure, imposing dependence on security measures offered by their CSP to stratagem and protection of their cloud installations. Unfamiliarity of the many businesses in protecting cloud infrastructure and often deployed many clouds with its own set of vendor-specific security controls leads a misconfiguration or security omission. This is another reason for exposure of organization's cloud-based resources to attackers.

Unauthorized Access: The public internet is used for accessing the cloud infrastructure installations, which is accessible outside the network boundary. This makes it easier for consumers and employees to use cloud infrastructure, while also making it easier for an attacker to get unauthorised access to an organization's cloud-based resources. Through poorly set security or compromised credentials, an attacker can gain direct access to the system, all without the organization's knowledge.

Insecure Interfaces/APIs: The customers are provided with variety of interfaces and application programming interface by the CSPs. Well-versed documentation of this also provided to the customers to make it easy of use. Hence there is need protecting the cloud-based infrastructure interfaces to avoid the misuse. A cyber attacker can utilise the customer documentation to reveal and exploit possible tactics for gaining access to and retrieving sensitive data from the organization's cloud infrastructure.

Hijacking of Accounts: Password is widely used for authentication purpose, when password is used for authentication, it should be strong enough. Password security normally becomes weak, since people use extremely weak password or repetition in passwords for authenticating more accounts. Stolen password becomes the cause for phishing attack and data breaches. In the second case, the repetition of passwords intensifies the attacks. As organizations run their critical business processes depending on cloud-based infrastructure and applications, one of the major security challenges is account hijacking. An attacker who gains access to an employee's credentials could gain access to important data or functionality, whereas a customer's credentials allow them complete control over their online account. Furthermore, as compared to on-premise infrastructure, cloud-based organisations' ability to respond to attacks may be less effective.

Lack of Visibility: The cloud infrastructure and network are not owned by the organization; it is outside of organizations network. This results in ineffectiveness of the many of typical network visibility solutions in cloud setting and lack of cloud specific security capabilities, which obstruct the organizations capability to screen and protect its cloud-based resources contrary to attacks.

External Sharing of Data: To make data sharing easy on cloud, collaboration is made available by sending an email invitation or by providing link to the shared resources, in this case anybody having the URL can view the shared resource. In case of link based sharing it is difficult to limit the access, still it is popular because it is easy compared to collaboration of intended by invitation. This causes a big security concern in cloud. In link-based sharing, the cyber attacker hacks the link transmitted or it is guessed. This stolen link is used for the unauthorised access to the shared resource. Furthermore, constraining the access to one recipient is difficult in link-based sharing.

Malicious Insiders: Some current or former employees (Insider) are given access to the organization network and its most sensitive data, sometimes these employees become major threat to the organization. For an unprepared organization, it is very difficult recognize the malicious insider. In cloud deployments, organizations cannot have control over the cloud infrastructure. Various typical security methods are useless in cloud installations due to the lack of control over the underlying infrastructure. Moreover, public internet is used to access cloud-based infrastructure and it also suffers frequent security misconfiguration. The above two reasons make more difficult to identify a malicious insider.

Cyberattacks: Cyber attackers made Cybercrime as a business, and they might choose their target by calculating the profit they can get out of it. The best choice is cloud-based infrastructure which is directly manageable from the public Internet, is often insecure, and contains a lot of sensitive information. Moreover, wide range of business uses the cloud, a successful attack can be replicated with greater possibility of success. This results in more frequent attacks on cloud environment.

Denial of Service Attacks: Various services of the cloud is utilised by many of the organizations. An attack which denies these services (Denial of Service (DoS) attack) to the organization will have vital impact on the business. The DoS attacks from multiple sources are done using the vulnerable machine in the internet and also uses the virtual machines in the cloud to create a more dangerous form DoS attack called Distributed Denial of Service (DDoS) attack.

Conclusion

The cloud computing paradigm and its characteristics makes it widespread use in the business organization. It provides various types of services in order to meet the various business needs. The infrastructure maintenance and the types of users leads different types of service models. Virtualization plays major role in the services provided by the cloud services, here hypervisor plays major role. Two types of hypervisors are employed in the cloud virtualization. Since, it is an internet-based service and hypervisor does not have the complete control of the hardware, utilities and the applications, the cloud services are prone to various kinds of attacks. In order to protect the various services from the attacker, security is a major concern in cloud computing. The effective mechanism which protects its users is a major challenge in cloud computing.

References

1. Mell, P. and Grance, T. 2011. The NIST Definition of Cloud Computing: Recommendations of the National Institute of Standards and Technology, U.S Department of Commerce, Special Publications 800-145. https://nvlpubs.nist.gov/nistpubs/legacy/sp/nistspecialpublication800-145.pdf.
2. Jin, H., Ibrahim, S., Bell, T., Gao, W., Huang, D. and Wu, S. 2010. Cloud types and services. In: Furht, B. and Escalante (eds) *Handbook of cloud computing.* Springer Science+Business Media, New York, USA. 335–355.
3. Akande, A.O., April, N.A. and Belle, J-P.V. 2013. Management Issues with Cloud Computing. In: *Proceedings of the Second International Conference on Innovative Computing and Cloud Computing*, ACM.
4. Rao, C.C., Leelarani, M. and Kumar, Y.R. 2013. Cloud: Computing Services and Deployment Models. *International Journal of Engineering and Computer Science 2*, 3389–3392.
5. Buyya, R., Vecchiola, C. and Selvi. S.T. 2013. Mastering Cloud Computing: Foundations and Applications Programming. Morgan Kaufmann, Elsevier Inc., Waltham, USA. Pp. 452.
6. Khurana, S. and Verma, A.G. 2013. Comparison of Cloud Computing Service Models: SaaS, PaaS, IaaS. *International Journal of Electronics & Communication Technology 4*, 29–32.
7. Singh, M. 2018. Virtualization in Cloud Computing – A Study. International Conference on Advances in Computing, Communication Control and Networking. 64–67. doi: 10.1109/ICACCCN.2018.8748398.
8. Jain, N. and Choudhary, S. 2016. Overview of virtualization in cloud computing. In: 2016 Symposium on Colossal Data Analysis and Networking. 1–4. doi: 10.1109/CDAN.2016.7570950.

9. Malhotra, L., Agarwal, D., and Jaiswal, A. 2014. Virtualization in Cloud Computing. *Journal of Information Technology & Software Engineering* **4**, 1–3.

10. Sharma, S., Singh, S., Singh, A. and Kaur, R. 2016. Virtualization in Cloud Computing. *International Journal of Scientific Research in Science, Engineering and Technology* **2**, 181–186.

11. Rashid, A. and Chaturvedi, A. 2019. Virtualization and its Role in Cloud Computing Environment. *International Journal of Computer Sciences and Engineering* **7**, 1131–1136.

12. Koganti, K.T., Patnala, E., Narasingu, S.S. and Chaitanya, J.N. 2013. Virtualization Technology in Cloud Computing Environment. *International Journal of Emerging Technology and Advanced Engineering* **3**, 771–773.

13. Xing, Y. and Zhan, Y. 2012. Virtualization and Cloud Computing. In: Zhang, Y. (eds) Future Wireless Networks and Information Systems. Lecture Notes in Electrical Engineering, vol 143. Springer, Berlin, Heidelberg. https://doi.org/10.1007/978-3-642-27323-0_39

CHAPTER 6

Mathematical Modeling and Analysis of Virally Infected Phytoplankton in Presence of Delay

Rakesh Kumar[1], Amanpreet Kaur[2], Krishna Pada Das[3]
[1]Department of Applied Science and Humanities, Shaheed Bhagat Singh State University, Ferozepur, Punjab, India
[2]Shaheed Bhagat Singh State University, Ferozepur, Punjab, India
[3]Department of Mathematics, Mahadevananda Mahavidyalaya, Barrackpore, Kolkata, India

1. Introduction

Phytoplankton are very small and drifting organisms but play a remarkable role in the world of the ocean. Phytoplankton are known as the primary source of the marine food web system and they are capable of reproducing nutrients and providing energy to the entire ocean ecosystem. Like other land plants, phytoplankton use chlorophyll to get sunlight and perform photosynthesis to get chemical energy. They consume carbon dioxide and produce oxygen for the world. Almost 50% of the oxygen that we breathe is generated by these phytoplankton. Growth of these phytoplankton is essential for the other marine species like zooplankton, fish etc. Phytoplankton is known as a rich nutrient source, but the quantity of oxygen that is required for the growth of marine species and human life can be affected with fast-growing phytoplankton. Harmful phytoplankton blooms due to some virus or disease. Virus in the ocean is considered as a mortality agent for phytoplankton and affects the survival and dynamics of other marine species too. Viruses have their different biological properties. The researcher has defined two types of viruses, 'Lysogenic' and 'Lytic' in phytoplankton and has discussed their nature. Viruses do not have their own metabolism and are dependent on other organisms for regrowth or any other energy requiring process. In lysogenic viruses, genes of the host organism are used for regrowing in the environment but in lytic virus infection, cell reproduction machinery of the host is used for regrowth [14]. In view of viral infections, the whole phytoplankton class has been classified into two categories- (a) Susceptible: who can be infected and (b) Infected: who are infected with some disease. It is also assumed that an infected population has less capacity to grow as compared to susceptible one. Dynamics of virally infected prey-predator models have been analysed by many researchers [3,4,7]. Author has given a comparative study of prey predator systems using different types of response functions for preying and explored that in linear response function, the system shows oscillations while stability has been described in type III response function. In addition to that, the author has analysed that the systematic changes in the size of the population held due to the natural surroundings [4]. The impact of viral infection on phytoplankton and zooplankton structures has also been analysed by many researchers and most of them have assumed that disease affects only prey population means phytoplankton [2,5,8,9]. Chattopadhyay [8] has explored that if viral infection communicates with law of mass of action, then a new infected class has been formed from the contact of virus to the susceptible class and pointed out that an instant change in the infection rate can disturb the stability of the structure. Some researchers have explained the toxin producing phytoplankton dynamics and their impacts on the growth of other species has also been discussed [10-12,17]. The author has studied the consequences of toxicity on dynamics of prey predator model when prey is infected with virus. Phytoplankton avoids becoming prey by zooplankton so tries to adopt the toxin production scheme to escape from predation. Phytoplankton produces toxins which diminishes the predation rate and consequently, the zooplankton population gets affected [16]. Generally, virus does not attack and reflect immediately; some time has been taken from the population of susceptible class to transformed into the infected class. This lag in time from subjection to the exposing of the infection is known as delay time, gestation or the incubation period. It plays an important role in transmission of infection. Sometimes due to immunity and incubated period, an incubated person can recover and become susceptible one [1,6,11,13-16]. Many researchers have taken the incubated class (the population which is exposed to be infected) in their study in place of the incubated period and observed that in the absence of the incubated class, the endemic equilibrium point showed un-stability but including the incubated class made it conditionally stable [14]. Contagious diseases cause substantial changes in the plankton system and the unexpected and complicated factors like rate of toxin generation and carrying capacity are capable of creating chaos in the plankton system whereas rate of infection transmission and recovery rate are capable of bringing the system into a stable position[16]. Delay period is necessary in studying the virally infected model for making an idea to know the causes and factors of infections and how it can be used for the better management of the system.

Taking the above concept under consideration, this paper will focus on the study of the effect of viral infections only on phytoplankton using 'delay period'. The paper is organised as section 2-virally infected phytoplankton population model with an incubated period has been developed and section 3- basic preliminaries including positivity, boundedness and equilibrium points have been discussed. Stability analysis of delay free model has been performed and by taking the bifurcation parameter as τ Hopf bifurcation has been investigated. Numerical simulation is being carried out to support our scientific findings in section 5 and at last the discussion section 6: our results of mathematical findings and their importance have been mentioned.

2. Formulation of the model

Effect of viral infections on the behaviour of phytoplankton dynamics, taking into account the 'delay period', is observed through a mathematical model. The population of phytoplankton has been classified into two categories such as 'susceptible' and 'infectious'. The basic preliminaries including positivity and boundedness have been discussed along with analysing the stability properties at endemic equilibrium (with and without delay). The Hopf bifurcation investigation proved that there exists a stable limit cycle by taking τ as a bifurcation parameter. Delay period in viral infection transformation plays a pivotal role in disappearance and stabilising the phytoplankton structure.

Here, we have assumed that the phytoplankton has been affected with some viral infection. The population of phytoplankton has been classified into two categories such as 'susceptible' $S(t)$ and 'infected' $I(t)$. It has been assumed that as compared to the susceptible ones, the infected class of phytoplankton regrows with less capability. We supposed that phytoplankton expand with r as intrinsic growth rate and K as carrying capacity. Susceptible population enters the infected population due to interaction of some virus. Rate of infection

89

with which virus contaminates following the law of mass of action has been taken as C. Keeping in view that some of the infected phytoplankton can recover and enter into the susceptible class, recovery rate has been assumed as γ [2,14,16]. The mass of infected phytoplankton increases when it interconnects with the susceptible phytoplankton and consequently, the susceptible population decreases. Mass of phytoplankton will diminish because of their natural death rate, so we assume susceptible class mortality rate δ_1 and δ_2 as the mortality rate of the infected ones. It is a well-known fact that viruses do not affect instantaneously, i.e., it takes little time to create infection in the susceptible population. The time to transfer a population from susceptible class to the infected class is called time lag and is taken as a delay period τ.

Taking into consideration all the above assumptions, virally infected delayed model has been formulated as below:

$$\frac{dS}{dt} = rS\left(1 - \frac{S+I}{K}\right) - CSI + \gamma I - \delta_1 S,$$

$$\frac{dI}{dt} = -\delta_2 I + CS(t-\tau)I(t-\tau) - \gamma I,$$

(2.1)

with initial condition,

$$\left(S(\theta), I(\theta)\right) = \left(\phi_1(\theta), \phi_2(\theta)\right) \in C\left([-\tau, 0], R_+^2\right), \phi_1(\theta) \geq 0, \phi_2(\theta) \geq 0, \theta \in [-\tau, 0]$$

where $C\left([-\tau, 0], R_+^2\right)$ represents the Banach space of continuous functions mapping the interval $[-\tau, 0]$ into R_+^2. The structure of virally infected phytoplankton in presence of delay period has been described in model (2.1). It is to be noticed here that for the existence of the susceptible class of phytoplankton which expands with r as intrinsic growth rate and diminishes with δ_1 as natural mortality rate, $r - \delta_1$ must be positive. It is explored here that the phytoplankton population is significantly affected with the delay period in presence of viruses.

3. Basic Results

In this section, we explore the possibilities of basic preliminaries of the model, i.e., positivity and boundedness and the equilibrium points of the system.

3.1 The positivity and boundedness of the system

Theorem 1. All solutions of the system are defined in the region

$$\Psi = \left\{ (S,I) \in R_+^2 : S+I = \frac{K}{4r\omega}(r+\omega)^2 + \varepsilon, \varepsilon > 0 \right\} \text{for suitably cho-}$$

sen value of $\omega \leq \delta_2$

Proof. Considering the model without delay period τ as

$$\frac{dS}{dt} = rS\left(1 - \frac{S+I}{K}\right) - CSI + \gamma I - \delta_1 S,$$

$$\frac{dI}{dt} = -\delta_2 I + CSI - \gamma I \tag{3.1}$$

To explore the boundedness and positivity of the system, Construct a new function as

$$X(t) = S(t) + I(t) \tag{3.2}$$

Differentiating X with respect to time t, we get

$$\frac{dX}{dt} = \frac{dS}{dt} + \frac{dI}{dt} = rS\left(1 - \frac{S+I}{K}\right) - \delta_1 S - \delta_2 I$$

$$\Rightarrow \frac{dX}{dt} \leq rS\left(1 - \frac{S}{K}\right) - \delta_2 I$$

Introducing positive numbers ω such that

$$\frac{dX}{dt} + \omega X \leq S\left[\omega + r\left(1 - \frac{S}{K}\right)\right] - (\delta_2 - \omega)I$$

$$\Rightarrow \quad \frac{dX}{dt} + \omega X \leq \frac{K}{4r}(r+\omega)^2 \tag{3.3}$$

Using the differential equality theorem for equation (3.3), we have

$$0 < X < \frac{K}{4r\omega}(r+\omega)^2\left(1 - e^{-\omega t}\right) + X_0 e^{-\omega t} \text{ where}$$

$$X_0 = X(0) = (S(0), I(0))$$

Thus $\lim\limits_{t \to \infty} Sup\, X(t) = \frac{K}{4r\omega}(r+\omega)^2$

Hence region $\Psi = \left\{(S,I) \in R_+^2 : S + I = \frac{K}{4r\omega}(r+\omega)^2 + \varepsilon, \varepsilon > 0\right\}$ con-

tains all the solutions of the system.

Therefore, positive and uniformly bounded solutions of virally infected models have been investigated.

3.2 Equilibrium analysis

The feasibility for the equilibrium of the model system (2.1) is given by

$$rS\left(1 - \frac{S+I}{K}\right) - CSI + \gamma I - \delta_1 S = 0,$$

$$-\delta_2 I + CSI - \gamma I = 0 \tag{3.4}$$

(a) The feasible trivial equilibrium is attained at $E^0 = (0,0)$.

(b) Infection free equilibrium $E^1 = \left(\frac{(r-\delta_1)K}{r}, 0\right)$ exists and it is

feasible as $r - \delta_1 > 0$

(c) On solving the equation (3.4), endemic equilibrium $E^* = \left(S^*, I^*\right)$ is obtained, where

$$S^* = \frac{\gamma + \delta_2}{C}$$

$$I^* = \frac{(\gamma + \delta_2)\left[(r - \delta_1)CK - r(\gamma + \delta_2)\right]}{C\left[r(\gamma + \delta_2) + CK\delta_2\right]}$$

$$= \frac{(\gamma + \delta_2)}{C}\left[\frac{(r - \delta_1)K - \dfrac{r(\gamma + \delta_2)}{C}}{\dfrac{r(\gamma + \delta_2)}{C} + K\delta_2}\right]$$

$$\Rightarrow I^* = S^* \frac{\left[(r - \delta_1)K - rS^*\right]}{\left[rS^* + K\delta_2\right]}$$

Hence, feasibility at endemic equilibrium point $E^* = (S^*, I^*)$ exist if

$$K > \frac{r(\gamma + \delta_2)}{C(r - \delta_1)}$$

4. Analysis of stability and Hopf bifurcation

All possibilities of stability around all the existing equilibrium points for the model without delay (3.1) and with delay (2.1) will be analysed in this section.

4.1 Local asymptotic stability of delay free model i.e., at $\tau = 0$.

The general variational matrix for the delay free model (3.1) has been computed as:

$$J(S, I) = \begin{pmatrix} r - \dfrac{2rS}{K} - \dfrac{rI}{K} - CI - \delta_1 & -\dfrac{rS}{K} - CS + \gamma \\ CI & -\delta_2 + CS - \gamma \end{pmatrix}$$

(a) The variational matrix at the trivial equilibrium point $E^0 = (0,0)$ is given by

$$J(E^0) = \begin{pmatrix} r - \delta_1 & \gamma \\ 0 & -\delta_2 - \gamma \end{pmatrix}$$

Its characteristic values are given by $(r - \delta_1, -\gamma - \delta_2)$

Therefore, un-stability has been seen around E^0 as one characteristic value $r - \delta_1 > 0$.

(b) The variational matrix at infection free equilibrium point $E^1 = \left(\dfrac{(r-\delta_1)K}{r}, 0 \right)$ is given by

$$J(E^1) = \begin{pmatrix} -r+\delta_1 & -r+\delta_1 - CK - \dfrac{CK\delta_1}{r} \\ 0 & -\delta_2 - \gamma + \dfrac{CK(r-\delta_1)}{r} \end{pmatrix}$$

At E^1, characteristic values are $\left(-r+\delta_1, -\delta_2 - \gamma + \dfrac{CK(r-\delta_1)}{r} \right)$.

Therefore, the equilibrium point E^1 is locally asymptotically stable if $K < \dfrac{r(\gamma + \delta_2)}{C(r-\delta_1)}$.

(c) Around the endemic equilibrium point $E^* = \left(S^*, I^* \right)$, the variational matrix is given as:

$$J(S^*, I^*) = \begin{pmatrix} -\dfrac{rS^*}{K} - \dfrac{\gamma I^*}{K} & -\dfrac{rS^*}{K} - \delta_2 \\ CI^* & 0 \end{pmatrix}$$

Its characteristic equation is $x^2 + a_1 x + a_2 = 0$ where

$$a_1 = \dfrac{rS^*}{K} + \dfrac{\gamma I^*}{K}, \quad a_2 = \left(\dfrac{rS^*}{K} + \delta_2 \right) CI^*$$

By Routh Hurwitz criteria, $E^* = \left(S^*, I^* \right)$ is locally asymptotically stable.

4.2 Stability analysis of delay model (2.1) at $E^* = \left(S^*, I^* \right)$

Stability for the delay model (2.1) at endemic equilibrium point $E^* = \left(S^*, I^* \right)$ will be analysed in this section. For this, we introduce the new variables $u(t) = S(t) - S^*, v(t) = I(t) - I^*$.

94

Therefore, the model transformed to

$$\frac{du}{dt} = a_{11}u(t) + a_{12}v(t),$$

$$\frac{dv}{dt} = a_{22}v(t) + b_{21}u(t-\tau) + b_{22}v(t-\tau). \qquad (4.1)$$

$$a_{11} = r - \frac{2rS^*}{K} - \frac{rI^*}{K} - CI^* - \delta_1 = \frac{-\gamma I}{S} - \frac{rS}{K},$$

$$a_{12} = -\frac{rS^*}{K} - CS^* + \gamma = -\frac{r(\gamma+\delta_2)}{CK} - \delta_2$$

$$a_{22} = -\gamma - \delta_2 \; b_{21} = CI^* \; b_{22} = CS^* = \gamma + \delta_2$$

$$\text{Thus } A = \begin{pmatrix} a_{11} & a_{12} \\ 0 & a_{22} \end{pmatrix} \text{ and } B = \begin{pmatrix} 0 & 0 \\ b_{21} & b_{22} \end{pmatrix}$$

The characteristic equation for the delay model (2.1) is

$$\left(A + e^{-\lambda\tau}B - \lambda I \right) = 0 \text{ i.e.,}$$

$$\begin{vmatrix} a_{11} - \lambda & a_{12} \\ e^{-\lambda\tau}b_{21} & a_{22} + e^{-\lambda\tau}b_{22} - \lambda \end{vmatrix} = 0$$

Thus, the characteristic equation is:

$$\left(a_{11} - \lambda \right)\left(a_{22} + e^{-\lambda\tau}b_{22} - \lambda \right) - e^{-\lambda\tau}a_{12}b_{21} = 0$$

$$\Rightarrow \lambda^2 - \lambda(a_{11} + a_{22}) + a_{11}a_{22} + e^{-\lambda\tau}(-\lambda b_{22} + a_{11}b_{22} - a_{12}b_{21}) = 0 \quad (4.2)$$

For having the stability of the model (2.1) at $E^* = \left(S^*, I^* \right)$, the equation (4.2) must have all eigenvalues with negative real parts but it is difficult to verify all infinite numbers of roots with negative real parts of this transcendental equation. So, to avoid the difficulty in locating the roots lying in the left half plane, we have used the concept of Hopf bifurcation, where the system loses its stability [19]. Taking the delay period τ as a bifurcation parameter, it is required to observe the value of τ for which the roots of equation (4.2) lie on the imaginary axis. If possible, say $\lambda = i\zeta$ is one of the purely imaginary roots of (4.2)

so

$$-\zeta^2 - i\zeta(a_{11} + a_{22}) + a_{11}a_{22} + (\cos\zeta\tau - i\sin\zeta\tau)(-i\zeta b_{22} + a_{11}b_{22} - a_{12}b_{21}) = 0$$

or $-\zeta^2 - i\zeta P + Q + (\cos\zeta\tau - i\sin\zeta\tau)(-i\zeta R + M) = 0$ \hfill (4.3)

Where $P = a_{11} + a_{22}$, $Q = a_{11}a_{22}$, $R = b_{22}$, $M = a_{11}b_{22} - a_{12}b_{21}$

Equating real and imaginary parts from equation (4.3), we have

$M\cos\zeta\tau - R\zeta\sin\zeta\tau = \zeta^2 - Q$ and $R\zeta\cos\zeta\tau + M\sin\zeta\tau = -\zeta P$

\hfill (4.4)

Eliminating delay parameter τ, we have

$$\zeta^4 + \zeta^2(P^2 - R^2 - 2Q) + Q^2 - M^2 = 0 \tag{4.5}$$

Or

$$v^2 + v(P^2 - R^2 - 2Q) + Q^2 - M^2 = 0, \text{ where } v = \zeta^2 \tag{4.6}$$

Here, it is easily verifiable that

$P^2 - R^2 - 2Q = (a_{11} + a_{22})^2 - (b_{22})^2 - 2a_{11}a_{22} = a_{11}^2 + a_{22}^2 - b_{22}^2 = a_{11}^2 > 0$

,

If $Q^2 - M^2 < 0$ the equation (4.6) and hence equation (4.5) has one

positive root say $\zeta_0^2 = v_0$.

Solving equation (4.4) for τ corresponding to value of ζ_0, we have

$$\tau_n = \frac{1}{\zeta_0}\cos^{-1}\left(\frac{M(\zeta_0^2 - Q) - \zeta_0^2 PR}{M^2 + R^2\zeta_0^2}\right) + \frac{2\pi n}{\zeta_0} \tag{4.7}$$

For $\tau = \tau_n$, the equation (4.2) has pair of roots lying on imaginary

axis as $\pm i\zeta_0$.

Define $\tau^* = \min\{\tau_n\}, n = 0,1,2,....$, smallest positive value of τ_n.

\hfill (4.8)

Therefore, at $E^* = (S^*, I^*)$ the endemic equilibrium point, Hopf bi-

furcation has been observed when $\tau = \tau^*$

Theorem 2. Transversality condition of the system (2.1) for the existence of Hopf bifurcation will occur at $\tau = \tau^*$ if $\left[\dfrac{d\operatorname{Re}\lambda}{d\tau}\right]_{\tau=\tau^*} > 0$ using [18], where τ^* is defined by the equation (4.8).

Proof. Rewriting the equation (4.2) as

$$\lambda^2 - \lambda P + Q + e^{-\lambda\tau}(-\lambda R + M) = 0 \tag{4.9}$$

Differentiating equation (4.9) w.r.t τ, we have

$$(2\lambda - P)\frac{d\lambda}{d\tau} + e^{-\lambda\tau}(-R)\frac{d\lambda}{d\tau} - \tau e^{-\lambda\tau}(-\lambda R + M)\frac{d\lambda}{d\tau} - \lambda e^{-\lambda\tau}(-\lambda R + M) = 0$$

It follows that
$$\left(\frac{d\lambda}{d\tau}\right)^{-1} = \frac{2\lambda - P - Re^{-\lambda\tau} - \tau e^{-\lambda\tau}(-\lambda R + M)}{\lambda e^{-\lambda\tau}(-\lambda R + M)}$$

$$= \frac{2\lambda - P}{-\lambda(\lambda^2 - \lambda P + Q)} - \frac{R}{\lambda(-\lambda R + M)} - \frac{\tau}{\lambda}$$

$$= \frac{\lambda^2 - Q}{-\lambda^2(\lambda^2 - \lambda P + Q)} - \frac{M}{\lambda^2(-\lambda R + M)} - \frac{\tau}{\lambda}$$

so

$$\left[\frac{d\,\mathrm{Re}\,\lambda}{d\tau}\right]_{\tau=\tau^*}^{-1} = \mathrm{Re}\left[\frac{\lambda^2 - Q}{-\lambda^2(\lambda^2 - \lambda P + Q)} - \frac{M}{\lambda^2(-\lambda R + M)} - \frac{\tau}{\lambda}\right]_{\tau=\tau^*}$$

$$\Rightarrow \left[\frac{d\,\mathrm{Re}\,\lambda}{d\tau}\right]_{\tau=\tau^*}^{-1} = \mathrm{Re}\left[\frac{-\zeta^2 - Q}{\zeta^2(-\zeta^2 - i\zeta P + Q)} - \frac{M}{-\zeta^2(-i\zeta R + M)} - \frac{\tau}{i\zeta}\right]$$

$$= \left[\frac{\zeta^4 - Q^2}{\zeta^2((Q - \zeta^2)^2 + \zeta^2 P^2)} + \frac{M^2}{\zeta^2(\zeta^2 R^2 + M^2)}\right]$$

$$= \left[\frac{\zeta^4 - Q^2}{\zeta^2(\zeta^2 R^2 + M^2)} + \frac{M^2}{\zeta^2(\zeta^2 R^2 + M^2)}\right] = \frac{\zeta^4 - (Q^2 - M^2)}{\zeta^2(\zeta^2 R^2 + M^2)} > 0,$$

as $Q^2 - M^2 < 0$.

Therefore, at $\tau = \tau^*$ the transversality condition has been satisfied for the existence of the Hopf bifurcation.

5. Numerical Simulations

The significance of the incubation period and the justification of our scientific findings obtained in the last sections for the delayed (2.1) and delay free model (3.1) will be explained in this section through numerical simulations. In the existence of viral infection and incubation period, to support our theoretical findings: time series and phase space graphs combined with the bifurcation diagrams has explored the behaviour of the phytoplankton system. To achieve our aim, we have taken following set of parameters as:

$$r = 35, \quad K = 250, \quad C = 0.07, \quad \delta_1 = 0.2, \quad \gamma = 0.07, \quad \delta_2 = 3.4$$

For these parameters, first, we explored the delay free model (3.1) to verify our findings and stability around the endemic equilibrium point E^* has been shown through time series and phase space graphs when $\tau = 0$ via Fig. 1. The System stability has been affected by increasing the value of the delay period τ which signifies the value of incubation period. For $\tau = 0.143$, Fig. 2 explains the existence of stable limit cycle around the endemic equilibrium point E^*, but at $\tau = 0.144$ the limit, cycle loses its stability clear from the phase portrait in Fig. 3. At $\tau = 0.3$ the system has an unstable limit cycle (Fig. 4). It has been described that the small change in time delay affects the population of phytoplankton. We have shown that after crossing the threshold value by the bifurcation parameter τ i.e., $\tau = 0.144 = \tau^*$: un-stability of the system has been observed.

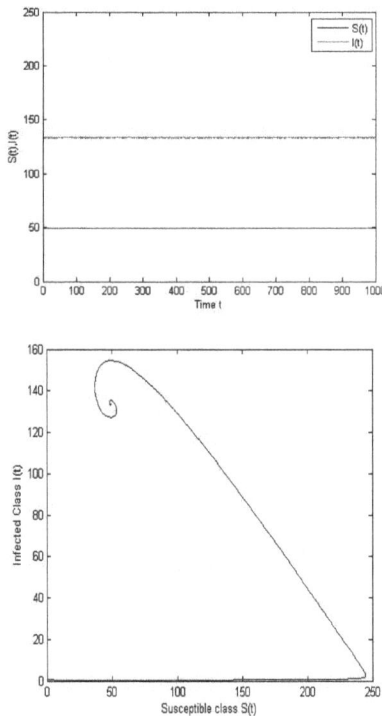

Fig. 1. Time series and phase space graphs show the convergence to stable equilibrium at $\tau = 0$

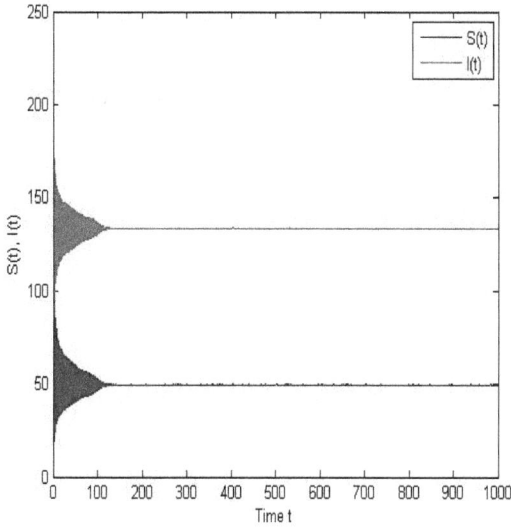

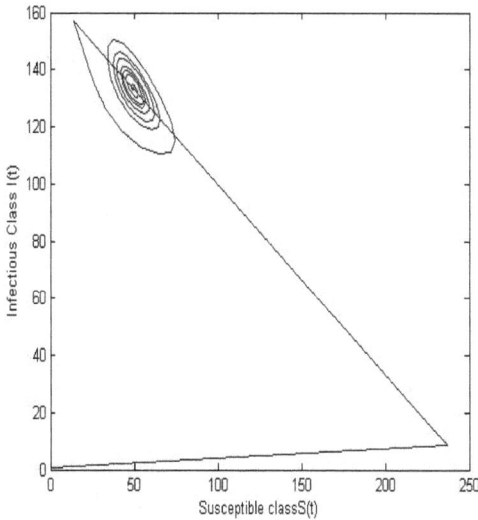

Fig. 2. Time series and phase space solution curves converging to stable equilibrium at $\tau = 0.143 < 0.144 = \tau^*$

To explore the long-term behaviour for different values of τ, we plot the bifurcation diagrams with respect to τ. As we increase the values of τ system induces its transition to instability. Figure 5 explains the stability of the system for the initial value of delay period τ. It has been observed that for small values of τ, the system is

99

stable but after crossing its threshold value $\tau^* = 0.144$ population of susceptible phytoplankton and infected phytoplankton bifurcates. It is noticeable that extending the value of the delay period strengthens the susceptible phytoplankton class but also weakens the infected phytoplankton class.

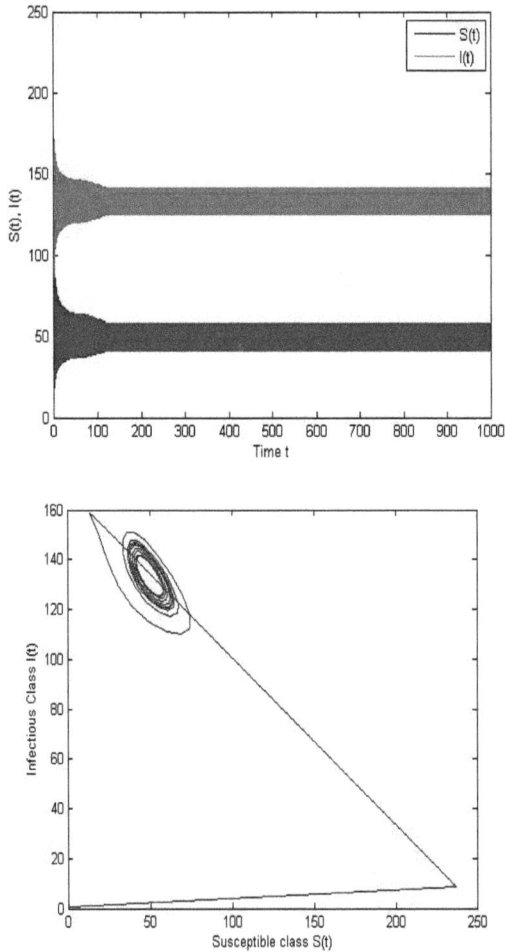

Fig. 3. Time series and phase portrait solution curves shows unstable equilibrium at $\tau = 0.144 = \tau^*$.

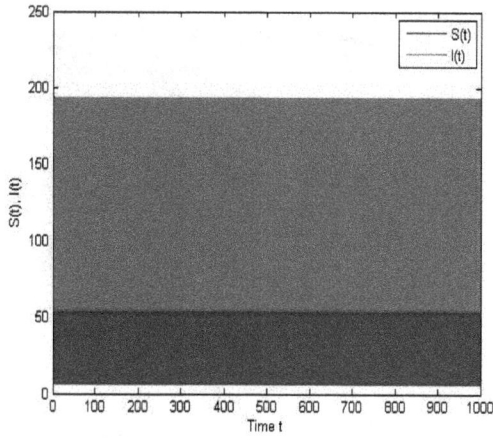

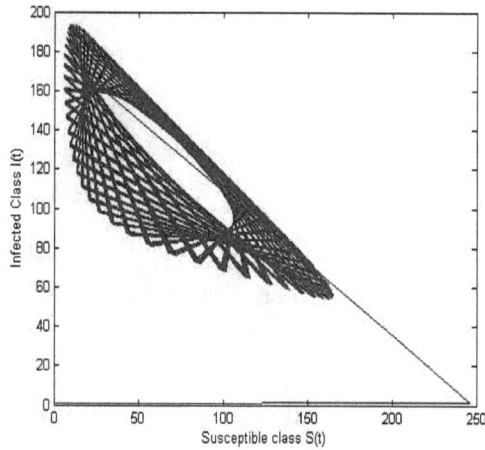

Fig. 4. Time series and phase portrait solution curves show unstable scenario at $\tau = 0.3 > \tau^*$

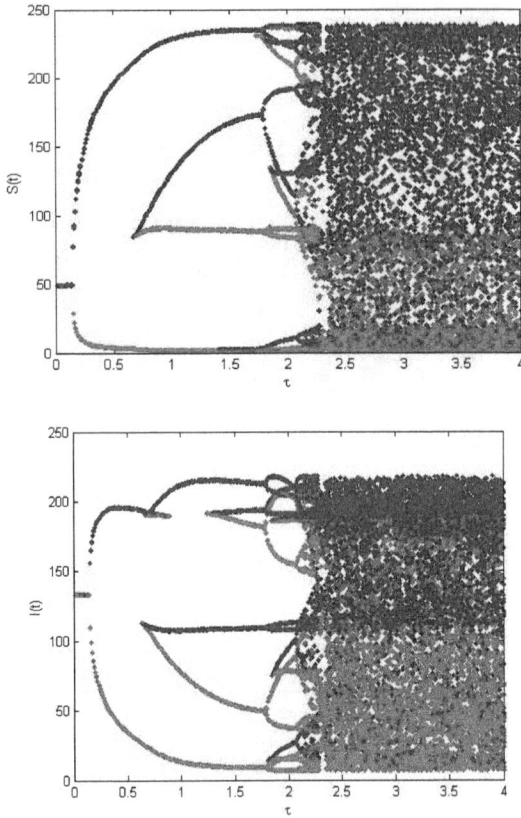

Fig. 5. Bifurcation diagrams show the dynamics of both classes of phytoplankton -susceptible and infected in appearance of 'delay period'.

5. Discussion

In existence of the delay period, a virally infected phytoplankton mathematical model has been investigated in this paper. To explore the dynamical changes in the phytoplankton with the length of gestation period, first we examined the delay free model and stability around the endemic equilibrium has been observed. In the delayed model, Hopf bifurcation has been examined with a critical parameter as delay period. Whenever delay period is $0 \leq \tau < 0.144$ system shows the static behaviour but after crossing its threshold value $\tau = 0.144$ (a very small value) system shows chaotic behaviour and the same has been described in the bifurcation diagrams of Fig. 5. Thus, there is a range of incubation time within which initially pursue stability then follows instability and finally, it leads to chaotic behaviour.

References

[1] Sharma, A., Sharma, A. K., & Agnihotri, K. (2014). The dynamic of plankton–nutrient interaction with delay. *Applied Mathematics and Computation*, *231*, 503-515.

[2] Agnihotri, K., & Kaur, H. (2019). The dynamics of viral infection in toxin producing phytoplankton and zooplankton systems with time delay. *Chaos, Solitons & Fractals*, *118*, 122-133.

[3] Auger, P., Mchich, R., Chowdhury, T., Sallet, G., Tchuente, M., & Chattopadhyay, J. (2009). Effects of a disease affecting a predator on the dynamics of a predator–prey system. *Journal of theoretical biology*, *258*(3), 344-351.

[4] Bairagi, N., Roy, P. K., & Chattopadhyay, J. (2007). Role of infection on the stability of a predator–prey system with several response functions—a comparative study. *Journal of Theoretical Biology*, *248*(1), 10-25.

[5] Singh, B. K., Chattopadhyay, J., & Sinha, S. (2004). The role of virus infection in a simple phytoplankton zooplankton system. *Journal of theoretical biology*, *231*(2), 153-166.

[6] Biswas, S., Saifuddin, M., Sasmal, S. K., Samanta, S., Pal, N., Ababneh, F., & Chattopadhyay, J. (2016). A delayed prey–predator system with prey subject to the strong Allee effect and disease. *Nonlinear Dynamics*, *84*(3), 1569-1594.

[7] Chattopadhyay, J., &Arino, O. (1999). A predator-prey model with disease in the prey. *Nonlinear analysis*, *36*, 747-766.

[8] Chattopadhyay, J., & Pal, S. (2002). Viral infection on phytoplankton–zooplankton system a mathematical model. *Ecological Modelling*, *151*(1), 15-28.

[9] Das, K. P., Roy, P., Karmakar, P., & Sarkar, S. (2020). Role of viral infection in controlling planktonic blooms-conclusion drawn from a mathematical model of phytoplankton-zooplankton system. *Differential Equations and Dynamical Systems*, *28*(2), 381-400.

[10] Gakkhar, S., & Negi, K. (2006). A mathematical model for viral infection in toxin producing phytoplankton and zooplankton systems. *Applied mathematics and computation*, *179*(1), 301-313.

[11] Gakkhar, S., & Singh, A. (2010). A delay model for viral infection in toxin producing phytoplankton and zooplankton system. *Communications in Nonlinear Science and Numerical Simulation*, *15*(11), 3607-3620.

[12] Khare, S., Misra, O. P., & Dhar, J. (2010). Role of toxin producing phytoplankton on a plankton ecosystem. *Nonlinear Analysis: Hybrid Systems*, *4*(3), 496-502.

[13] Dhar, J., & Sharma, A. K. (2009). The role of the incubation period in a disease model. *Applied Mathematics E-Notes*, *9*, 146-153.

[14] Dhar, J., & Sharma, A. K. (2010). The role of viral infection in phytoplankton dynamics with the inclusion of incubation class. *Nonlinear Analysis: Hybrid Systems*, *4*(1), 9-15.

[15] Dhar, J., Sharma, A. K., Sahu, G. P., & Bhatti, H. S. (2011). Mathematical modelling and analysis of viral disease outbreak with partial immunity and incubation period. *Elixir Bio. Phys*, *37*, 3691-3695.

[16] Thakur, N. K., Srivastava, S. C., & Ojha, A. (2021). Dynamical study of an eco-epidemiological delay model for plankton systems with toxicity. *Iranian Journal of Science and Technology, Transactions A: Science*, *45*(1), 283-304.

[17] Upadhyay, R. K., & Chattopadhyay, J. (2005). Chaos to order: role of toxin producing phytoplankton in aquatic systems. *Nonlinear Analysis: Modelling and Control*, *10*(4), 383-396.

[18] Liu, J., Lv, P., Liu, B., & Zhang, T. (2021). Dynamics of a Predator-Prey Model with Fear Effect and Time Delay. *Complexity*, *2021*.

[19] Ojha, A., & Thakur, N. K. (2020). Exploring the complexity and chaotic behaviour in plankton–fish system with mutual interference and time delay. *Bio Systems*, *198*, 104283.

[20] Zhou, X., Shi, X., & Song, X. (2009). Analysis of a delayed prey-predator model with disease in the prey species only. *Journal of the Korean Mathematical Society*, *46*(4), 713-731.

[21] Rehim, M., & Imran, M. (2012). Dynamical analysis of a delay model of phytoplankton–zooplankton interaction. *Applied Mathematical Modelling*, *36*(2), 638-647.

[22] Huang, C., Zhang, H., Cao, J., & Hu, H. (2019). Stability and Hopf bifurcation of a delayed prey–predator model with disease in the predator. *International Journal of Bifurcation and Chaos*, *29*(07), 1950091.

[23] Wang, S., Song, X., & Ge, Z. (2011). Dynamics analysis of a delayed viral infection model with immune impairment. *Applied Mathematical Modelling*, *35*(10), 4877-4885.

[24] Al Basir, F., Tiwari, P. K., & Samanta, S. (2021). Effects of incubation and gestation periods in a prey–predator model with infection in prey. *Mathematics and Computers in Simulation, **190**,* 449-473.

CHAPTER 7

New Advances in Wavelet Theory

Meenakshi
Mathematics, LRDAV College, Jagraon, Punjab, India

1. Introduction

Wavelet is a recent advancement having enrichment of mathematical content and great potential for applications. Wavelet analysis is simple and versatile tool in applied mathematics having the important applications in different fields. It is well formalized and a vital area of research in different origins. This theory is a blend of various ideas which derived from different disciplines which includes mathematics (Littlewood-Paley theory and Calderon- Zygmund operators), engineering (quadrature mirror filters, subband encrypting in signal transmission and pyramidal designs in imaging) and physics (renormalization group and in quantum mechanics, the study of Coherent states mechanism). The wavelet theory has proved a great role in integration of different streams such as physics, mathematics and engineering. In the recent few years wavelets are developed by practical deliberations in mathematics, engineering and in different fields.

In mathematical and engineering analysis methods are required to represent a compound data and algorithms in terms of simple, well understood algorithms. In this context wavelet theory has proved to be useful which has given new methods for decomposition of a function or a signal. It has provided an approach in analyzing time-varying, non-stationary real-world signals. A new class of orthogonal expansions in $L^2(R)$ has been provided with good time-frequency analysis and uniformity estimation properties, see [5, 22, 23]. Solution of partial differential equations [7, 10], modeling real world problems, signal and image processing, analysis of signals in biomedical sciences [1-3, 7, 8, 12] have been studied through wavelets. It has also applications in oil industry, meteorology and in analyzing financial time series [12, 15].

Fourier series and transforms is the marvellous discovery in the mathematics and it has important applications in different fields such as mathematics, physics and engineering. But around 1940's its shortcomings were realized by scientists [5]. Time localization of the spectral component is not presented by Fourier [5]. Time-frequency representation(TFR) is required in the analyzing inconsistent waves whose spectral content changes with time rather than only a frequency representation[11].

Wavelets in the present theoretical form were introduced by Grossman and Morlet as an important tool for signal analysis at the beginning of the nineteen eighties [13] for the analysis of seismic data. They considered the family of functions of

$$\varphi_{a,b}(x) = |a|^{-1/2} \varphi\left(\frac{x-b}{a}\right), a, b \in R, a \neq 0$$

translated and dilated versions of a function with scaling parameter a which estimates the compression or scale and translation parameter b. Wavelet analysis consists in applying such families of functions to decompose data, functions or operators. Alex Grossman, a French theoretical physicist, evolved a transposition formula for wavelet transform on recognizing the importance of Morlet transforms.

Followed by introduction to wavelets in section 1, in section 2 a brief review of Fourier and Wavelet theory has been given. In section 3 the study of multiresolution analysis is explained with brief examples. The process of decomposition and reconstruction of wavelets is explained in section 4. Haar-Vilenkin wavelets are studied in section 5 and it is devoted to study of various properties of Haar-Vilenkin wavelets.

2 From Fourier analysis to Wavelets

2.1 Study of Fourier Series and Transforms

Fourier analysis is the classical area of mathematics which deals with the representation of functions through trigonometric functions on R.

Consider an infinite series of the form

½ a₀+ (a₁ cos x + b₁ sin x)+ (a₂ cos 2x + b₂ sin 2x) + ...

which can be written as

$$\frac{1}{2}a_0 + \sum_{k=1}^{\infty}(a_n \ coskx + b_n sinkx). \tag{2.1}$$

It is called a trigonometric series.

On $[-\pi, \pi]$ consider a piecewise continuous function f. Consider the numbers

$$a_0 = \frac{1}{\pi}\int_{-\pi}^{\pi} f(x)dx, \quad a_k = \frac{1}{\pi}\int_{-\pi}^{\pi} f(x) \ coskxdx, \tag{2.2}$$

$$b_k = \frac{1}{\pi}\int_{-\pi}^{\pi} f(x) \ sinkx \ dx. \tag{2.3}$$

For k = 1,2,3,...and on $[-\pi, \pi]$, these numbers are called Fourier sine and cosine coefficients of a function f and the series (2.1) having coefficients of the type given by equations (2.2) and (2.3) is Fourier series of f.

When we represent $f: [-\pi, \pi] \to R$ in a trigonometric series of the type,

$$\frac{1}{2}a_0 + \sum_{k=1}^{\infty}(a_k \ coskx + b_k sinkx), \tag{2.4}$$

we get the Fourier coefficients. Consider the partial sums for an arbitrary trigonometric series which are defined as

$$s_N(x) = \frac{1}{2}a_0 + \sum_{k=1}^{N}(a_n \ coskx + b_n sinkx). \tag{2.5}$$

Using the orthonormality of the system {cosnx, sinnx, n= 1, 2, 3...} on $[-\pi, \pi]$, for $N \geq n \geq 1$, let us calculate

$$\int_{-\pi}^{\pi} s_N(x)cosnx \ dx = \frac{1}{2}a_0 \int_{-\pi}^{\pi} cosnx \ dx +$$
$$\sum_{k=1}^{N} a_k \int_{-\pi}^{\pi} cosnx \ coskx \ dx + \sum_{k=1}^{N} b_k \int_{-\pi}^{\pi} cosnx \ dx \ sinkx \tag{2.6}$$

$$= 0 + a_n \pi + 0 = \pi \, a_n.$$

From equation (2.4), we have $f(x) = \lim_{N \to \infty} s_N(x)$, and

$$\int_{-\pi}^{\pi} f(x) \cos kx \, dx = \lim_{N \to \infty} \int_{-\pi}^{\pi} s_N(x) \cos kx \, dx \, .$$

Then from equation (2.6), we have

$$a_k = \frac{1}{\pi} \int_{-\pi}^{\pi} f(x) \cos kx dx.$$

Similarly, on using sin kx instead of cos kx in (2.6), we obtain the value of b_k. We can consider the interval [-l,l] in place of $[-\pi, \pi]$ for an arbitrary number l where l>0.

Definition 2.1 (Fourier series)

For l >0, on the interval [-l, l], consider function f as a piecewise continuous. Then on   [-l,l], f has a Fourier series as

$$\frac{1}{2}a_0 + \sum_{k=1}^{\infty}(a_k \, \cos\frac{k\pi x}{l} + b_k \sin\frac{k\pi x}{l}), \ x \in [-l, l].$$

where
$$a_0 = \frac{1}{l}\int_{-l}^{l} f(x)dx, \ a_k = \frac{1}{l}\int_{-l}^{l} f(x) \cos\frac{k\pi x}{l} dx, \tag{2.7}$$

$$b_k = \frac{1}{l}\int_{-l}^{l} f(x) \sin\frac{k\pi x}{l} \, dx. \tag{2.8}$$

If the above series converges to f(x) at x, we write it as

$$f(x) = \frac{1}{2}a_0 + \sum_{k=1}^{\infty}(a_k \, \cos\frac{k\pi x}{l} + b_k \sin\frac{k\pi x}{l}).$$
$$\tag{2.9}$$

Fourier series has provided us a powerful tool for representation of periodic functions. For the representation of aperiodic functions we have a different tool that is Fourier transform.

Definition 2.2 (Fourier transform)

The Fourier transform of f(x) where x is taken on R and also integrable on R denoted by

$$\hat{f}(\theta) = \int_{-\infty}^{\infty} f(t)e^{-ti\theta}\, dt .$$ (2.10)

Fourier theory is a central ingredient of signal theory and representation. From a given signal it is used to extract the exact information. This theory is used where it is necessary to analyze the frequency of a signal. It is useful in the study of signals which consist of several oscillations with different amplitudes for example music signals and signals of a engine. The modern supplement to classical Fourier analysis is wavelet analysis. Fourier analysis only gives the spectral content and not represents the time localization of frequency content, see [5]. Wavelet theory has proved useful in this analysis. Wavelet provides the representation of the form

$$f(x) = \sum_{n=0}^{\infty} a_n f_n(x)$$

for a large class of functions f [6,10].

2.2 Wavelet Analysis

Definition 2.3 Wavelet

A wavelet is a small wave, and wavelet means an oscillation that decays quickly. Mathematically, a function ψ that meets the conditions written below is called a wavelet:

(i) Energy defined by E of ψ must be finite, i.e. $E =$
$$\int_{-\infty}^{\infty} |\psi(t)|^2\, dt < \infty .$$ (2.11)

(ii) The following condition must hold:
$$C_\psi = \int_{-\infty}^{\infty} \frac{|\psi(\xi)|^2}{|\xi|}\, d\xi < \infty$$

(2.12)

Here C_ψ is the admissibility constant and the above condition is the admissibility condition and.

(iii)   The function ψ has a zero mean, i.e. $\hat{\psi}(0) = 0$.

Definition 2.4 For the transformation of a signal into another representation, the wavelet methods are used which provides the useful information about a signal. This transform is called a wavelet transform. For f on R which is square integrable, the continuous wavelet transform is denoted by T_ψ with the wavelet, ψ is defined as

$$T_\psi f(a,b) = a^{-1/2} \int_{-\infty}^{\infty} f(t)\psi^* \left(\tfrac{t-b}{a}\right) dt, \qquad (2.13)$$

where $a \in R - \{0\}, b \in R$ and ψ^* denotes the complex conjugate of ψ. For a real-valued function $\psi^* = \psi$. Here a denotes the scale or frequency parameter and b denotes the translation parameter. Since $\psi_{a,b}(t) = \frac{1}{\sqrt{a}}\psi\left(\frac{t-b}{a}\right)$, therefore

$$T_\psi f(a,b) = \int_{-\infty}^{\infty} f(t)\psi_{a,b}{}^*(t)\, dt,$$

i.e.

$$T_\psi f(a,b) = <f, \psi_{a,b}>.$$

Definition 2.5 The **inverse wavelet transform** is defined as

$$f(t) = \frac{1}{C_\psi}\int_{-\infty}^{\infty}\int_{0}^{\infty} T_\psi f(a,b)\,\psi_{a,b}(t)\,\frac{da\,db}{a^2}, \qquad (2.14)$$

With the help of inverse transform by integrating wavelet transform over the scale and translation parameter, the original signal is recovered.

Definition 2.6 A wavelet $\psi \in L^2(R)$ is an orthonormal wavelet if the family $\{\psi_{j,k},\ j,k \in Z\}$, where

$$\psi_{j,k(x)} = 2^{\frac{j}{2}}\psi(2^j x - k), j, k \in Z,$$

in $L^2(R)$ constitute an orthonormal basis i.e.

$$<\psi_{j,k}, \psi_{l,m}> = \delta_{j,k}\delta_{l,m}, \qquad j,k,l,m \in Z$$

112

Where $\delta_{j,k}$ denotes the Kronecker's delta and every $f \in L^2(R)$ can be written as

$$f(x) = \sum_{j \in Z} \sum_{k \in Z} d_{j,k}\, \psi_{j,k}(x),$$

Where $d_{j,k}$ denotes the sequence of scalars.

Definition 2.7 The wavelet coefficients of $f \in L^2(R)$ are defined as
$$d_{j,k} = <f, \psi_{j,k}>, j, k \in Z.$$

The wavelet series of $f \in L^2(R)$ is $\sum_{j \in Z} \sum_{k \in Z} <f, \psi_{j,k}> \psi_{j,k}(x)$.

The wavelet representation of f is

$$f(x) = \sum_{j \in Z} \sum_{k \in Z} <f, \psi_{j,k}> \psi_{j,k}(x).$$

The amount of fluctuation about the point $t = 2^{-j}k$ is estimated by the point $d_{j,k}$ with the frequency determined by dilation index j. Also wavelet coefficient at point $(2^{-j}, k2^{-j})$ is

$$d_{j,k} = T_\psi f(2^{-j}, k2^{-j}).$$

Examples of Wavelet

Example 2.1: (Haar Wavelet) The first wavelet ever constructed was Haar wavelet, see Figure 1. The simplest example of wavelet which is orthonormal is Haar wavelet, see [14] which is taken as:

$$\psi(t) = \begin{cases} 1 & 0 \le t < \tfrac{1}{2} \\ -1 & \dfrac{1}{2} \le t \le 1 \\ 0 & otherwise \end{cases}$$

And Haar scaling function is

$$\phi(x) = \begin{cases} 1, & 0 \ll t < 1 \\ 0, & otherwise \end{cases}$$

i.e. $\phi(t) = \phi(2t) + \phi(2t-1)$.

The corresponding Haar wavelet equation is $\psi(t) = \phi(2t) + \phi(2t-1)$.

Figure 1

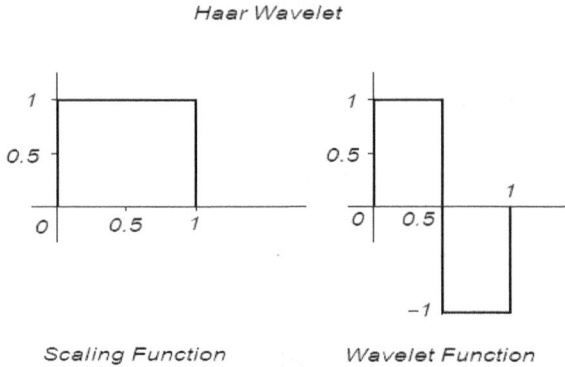

Haar Wavelet

Scaling Function   Wavelet Function

Example 2.2 (Mexican Hat Wavelet) It is a function denoted by ϕ is defined as $\phi(x) = (1-x^2)e^{-x^2/2}$ is known as Mexican Hat Wavelet. For Gaussian Distribution function $e^{-x^2/2}$ this wavelet is the negative of second derivative of it.

It has no scaling function and no discontinuities.

3. Multiresolution Analysis

Multiresolution analysis is a heart of wavelet theory and provides a tool for decomposition and reconstruction of wavelets. Multiresolution means that we want to use different degrees of resolution in order to approximate a given function f, ranging from coarse to fine structures, like a camera zooming in on an object. Mathematically these are defined as different degrees of resolution by a nested sequence of vector spaces of functions,

$$\_ \subset V_{-2} \subset V_{-1} \subset V_0 \subset V_1 \subset V_2 \subset \_ \tag{3.1}$$

Definition 3.1: A one dimensional Multiresolution analysis (abbreviated MRA) in 1989 under the directions of Meyer was introduced by Mallat [16]. It is a sequence of subspaces $\{V_j\}_{j\in Z}$ which are closed and are square integrable over R and the sequence is increasing such that

(i) $\bigcap_{j\in Z} V_j = \{0\}$
(ii) is dense in $L^2(R)$.
(iii) $\bigcup_{j\in Z} V_j$ $f(x) \in V_j$ iff $f(2x) \in V_{j+1}$ and if $f(x-m) \in V_j$ for $m \in Z$.
(iv) There exists a square integrable function $\phi(x)$ over R called as a scaling function such that the collection $\{\phi(x-n)\}$ of translates where $n \in Z$ is an orthonormal system and $V_0 = \overline{span}\{\phi(x-n)\}$.

With the given sequence of MRA V_j and the scaling function ϕ, then a wavelet is constructed by:

Suppose the subspace W_j of $L^2(R)$ is defined as

$$V_j \oplus W_j = V_{j+1}, \; V_j \perp W_j \forall j.$$

We define $J_j(f(x)) = f(2^j x) \forall j \in Z$. Thus $J_j(V_1) = V_{j+1}$.

Thus
$$V_j \oplus W_j = J_j(V_0 \oplus W_0) = J_j(V_0) \oplus J_j(W_0) = V_j \oplus J_j(W_0).$$
Then
$$J_j(W_0) = W_j \forall j \in Z. \tag{3.1}$$

Using the conditions (i) and (ii) of Definition 3.1, we have the orthogonal decomposition
$$\oplus_{j\in Z} W_j = L^2(R). \tag{3.2}$$
Thus it is required to obtain $\phi \in W_0$ such that it's system of translates of over Z forms an orthonormal basis in W_0. By the equations (3.1) and (3.2) such function is a wavelet function. The wavelet obtained from MRA is called a wavelet associated with MRA.

115

Remark 3.1: For defining MRA, we first find the subspace V_0 by taking V_j as

$$V_j = \{f(x): f(x) = D_{2^j}g(x) \text{ where } g(x) \in V_0\},$$

so that the first part of condition (iii) of Definition 3.1 is satisfied and then we prove that the Definition 3.1(i), (ii), (iv) and equation 3.1 hold. First, we identify the function ϕ such that translates of the function over Z i.e. $\{\phi(x-m): m \in Z\}$ is an orthonormal system, after this, we will define $V_0 = \overline{span\{\phi(x-n)\}}$.

Definition 3.2 Take

$$\phi_{j,k}(x) = 2^{\frac{j}{2}}\phi(2^j x - k) \,\forall\, j, k \in Z.$$

We define the **approximation operator** of $f \in L^2(R)$ denoted by P_j as

$$P_j f(x) = \sum_k <f, \phi_{j,k}> \phi_{j,k}(x) \forall\, j \in Z.$$

The **detail operator** of $f \in L^2(R)$ denoted by Q_j is defined as

$$Q_j f(x) = P_{j+1}f(x) - P_j f(x).$$

Lemma 3.1 $\{\phi_{j,k}, j, k \in Z\}$ forms orthonormal basis for $V_j \,\forall\, j \in Z$.

For proof, we refer to [22].

Lemma 3.2 For all f, C_c^0 on R

  (i) $\lim\limits_{j\to\infty}\|P_j f - f\|_2 = 0$ and

  (ii) $\lim\limits_{j\to-\infty}\|P_j f\|_2 = 0$.

Proof (i) Let $\varepsilon > 0$ be arbitrary. For $A \in Z$ there exists $g(x) \in V_A$ by using definition 3.1

(ii), we have $\|f - g\|_2 < \varepsilon/2$.

By Definition 3.1(i), $g(x) \in V_A$ and $P_a g(x) = g(x) \forall\, a \geq A$. Hence by using Minkowski and Bessel's inequality, we have

$$\|f - P_a f\|_2 = \|f - g + P_a g - P_a f\|_2$$
$$\leq \|f - g\|_2 + \|P_a(f - g)\|_2$$
$$\leq \|f - g\|_2 < \varepsilon.$$

Since the result is true $\forall\, a \geq A$, hence the result is proved.

  (ii) Let $\varepsilon > 0$ be given and assume that f(x) is supported on the interval $[-\alpha, \alpha]$. By using Cauchy-Schwarz and Minkowski's inequality and using the orthonormality of $\{\gamma_{a,b}\}_{b\in Z}$, we have

$$\|P_a f\|_2^2 = \left\|\sum_b < f, \gamma_{a,b} > \gamma_{a,b}\right\|_2^2$$

$$= \sum_b | < f, \gamma_{a,b} > |^2$$

$$= \sum_b | \int_{-\alpha}^{\alpha} 2^{\frac{a}{2}} f(x)\, \gamma(2^a x -$$

b) $dx|^2$

$$\leq$$

$$\sum_b \left(\int_{-\alpha}^{\alpha} |f(x)|^2 dx\right) 2^a \left(\int_{-\alpha}^{\alpha} |\gamma(2^a x - b)|^2\, dx\right)$$

$$=$$

$$\|f\|_2^2 \sum_b \int_{-2^a\alpha-b}^{2^a\alpha-b} |\gamma(x)|^2\, dx .$$

We are required to prove that

$$\lim_{a\to-\infty} \sum_b \int_{-2^a\alpha-b}^{2^a\alpha-b} |\gamma(x)|^2\, dx = 0.$$

In order to prove it, consider $\epsilon > 0$ and select β so large such that

$$\sum_{|b|\geq\beta} \int_{-1/2-b}^{1/2-b} |\gamma(x)|^2\, dx = \int_{|x|\geq\beta-1/2} |\gamma(x)|^2\, dx < \varepsilon.$$

Also $\lim_{a\to-\infty} \int_{-2^a\alpha-b}^{2^a\alpha-b} |\gamma(x)|^2 dx = 0 \ \forall\ b \in Z$. Therefore,

$$\lim_{a\to-\infty} \|P_a f\|_2^2 \leq \|f\|_2^2 \lim_{a\to-\infty} \sum_b \int_{-2^a\alpha-b}^{2^a\alpha-b} |\gamma(x)|^2\, dx$$

$$= \|f\|_2^2 \lim_{a\to-\infty} \left(\sum_{|b|\leq\beta} \int_{-2^a\alpha-b}^{2^a\alpha-b} |\gamma(x)|^2 +\right.$$

$$\left.\sum_{|b|>\beta} \int_{-2^a\alpha-b}^{2^a\alpha-b} |\gamma(x)|^2 \right)$$

$$\leq \|f\|_2^2 \lim_{a\to-\infty} \left(\varepsilon + \sum_{|b|>\beta} \int_{-2^a\alpha-b}^{2^a\alpha-b} |\gamma(x)|^2 \right)$$

$$= \|f\|_2^2 \varepsilon.$$

Theorem 3.1 Prove that ϕ in $L^2(R)$ can be expressed as

$$\phi(x) = \sum_m h(m) 2^{\frac{1}{2}}\phi(2x - m), \tag{3.3}$$

where h(m) is a l^2 sequence of coefficients. Also

$$\hat{\phi}(\gamma) = m_0\left(\frac{\gamma}{2}\right)\hat{\phi}\left(\frac{\gamma}{2}\right), \tag{3.4}$$

where

$$m_0(\gamma) = \frac{1}{\sqrt{2}}\sum_m h(m)e^{-2\pi i m\gamma} . \tag{3.5}$$

Proof Since $\phi \in V_0 \subseteq V_1$ and $\{\phi_{1,m}\}_{m\in Z}$ constitute orthonormal basis for V_1,

$$\phi(x) = \sum_m < \phi, \phi_{1,m} > 2^{\frac{1}{2}}\phi(2x - m).$$

Thus equation (3.3) holds and $h(m) = < \phi, \phi_{1,m} >$, by using Bessel's inequality, the sequence is l^2. On taking the Fourier transform of equation (3.3), equation (3.4) holds.

Definition 3.3 Consider the scaling function ϕ associated with MRA $\{V_j\}$. From equation (3.3) the sequence $\{h(m)\}$is the **scaling filter** and the $m_0(\gamma)$ from (3.5) is the **auxiliary function** corresponding to $\phi(x)$.

The following theorem gives an algorithm for construction of wavelets corresponding to given MRA.

Theorem 3.2 Consider MRA $\{V_j\}$ with scaling filter $\{h(m)\}$ and scaling function ϕ. Then the wavelet filter denoted by $\{g(m)\}$ is defined as

$$g(m) = (-1)^m \overline{h(1 - m)}. \tag{3.6}$$

The wavelet ψ is defined as

$$\psi(x) = \textstyle\sum_m g(m) \, 2^{\frac{1}{2}}\phi(2x - m), \tag{3.7}$$

then $\{\psi_{j,k}\}_{j,k\in Z}$ is called a wavelet orthonormal basis on R.
For $J \in Z$,

$$\{\phi_{J,k}(x)\}_{k\in Z} \cup \{\psi_{j,k}(x)\}_{j,k\in Z}$$

is an orthonormal basis on R.

Remark 3.2 Since $V_1 = \overline{span}\{\psi_{1,k}(x)\}$, equation (3.7) implies that $\psi(x) \in V_1$. On taking Fourier transform on both sides of equation (3.7)

$$\hat{\psi}(\gamma) = m_1\left(\frac{\gamma}{2}\right)\hat{\phi}\left(\frac{\gamma}{2}\right),$$

where

$$m_1(\gamma) = \frac{1}{\sqrt{2}}\sum_n g(n)e^{-2\pi i \gamma n}.$$

Definition 3.4 The wavelet subspace W_j is defined as
$$W_j = \overline{span}\{\psi_{j,k}(x)\}_{k\in Z}.$$

Example 3.1 (Haar MRA)

Let V_0 be the space of all the functions which are square integrable over R and constant over all the intervals [k, k+1) for integer k. According to condition (iv) of Definition 3.1, V_j is taken as the space of all square-integrable functions which are constant on all intervals of the type [k.2⁻ʲ, (k+1).2⁻ʲ). For ϕ, we choose

$$\phi(x) = \chi_{[0,1)}(x) = \begin{cases} 1, & 0 \le x < 1 \\ 0, & otherwise \end{cases}.$$

Thus $(x) = \frac{1}{\sqrt{2}}\phi_{1,0}(x) + \frac{1}{\sqrt{2}}\phi_{1,1}(x)$.

Therefore

$$h(n) = \begin{cases} \frac{1}{\sqrt{2}} & if \ n=0,1 \\ 0 & if \ n \ne 0,1 \end{cases}.$$

From equation (3.6), we have

$$g(n) = \begin{cases} \frac{1}{\sqrt{2}} & if \ n=0 \\ -\frac{1}{\sqrt{2}} & if \ n=1 \\ 0 & if \ n \ne 0,1 \end{cases}.$$

Therefore by equation (3.7)
$\psi(x) = \frac{1}{\sqrt{2}}\phi_{1,0}(x) - \frac{1}{\sqrt{2}}\phi_{1,1}(x)$. Thus $\{\psi_{j,k}\}_{j,k \in z}$ is a wavelet orthonormal basis on R.

4. Wavelet decomposition and reconstruction algorithm

Consider the general structure of MRA where $\phi, \psi \in L^2(R)$ are the corresponding scalar and wavelet functions respectively and generate $\{V_j\}$ and $\{W_j\}$. B y Definition 3.1(ii), for $m \in Z$ the function $f \in L^2(R)$ can be approximated by $f_m \in V_m$. Also $V_m = V_{m-1} \oplus W_{m-1} \forall m \in Z$. Thus for $f_{m-1} \in V_{m-1}$ and $h_{m-1} \in W_{m-1}$ such that f_m has a decomposition of the type $f_m = f_{m-1} + h_{m-1}$. By using the process, again and again, we have

$$f_m = h_{m-1} + h_{m-2} + \ldots + h_{m-M} + f_{m-M}, \tag{4.1}$$

where for all $j \in Z$, $f_j \in V_j$ and $h_j \in W_j$.

The decomposition in (4.1), which is unique is a wavelet decomposition. In order to stop, it is required to $\|f_{m-M}\|$ be smaller than some threshold value.

Since both the scaling function $\phi \in V_0$ $\psi \in W_0$ lies in V_1 and the V_1 is generated by $\phi_{1,m}(x) = 2^{\frac{1}{2}}\phi(2x-m), m \in Z$, there correspond the sequences $\{p_k\}$ and $\{g_k\} \in l^2$ such that

$$\phi(x) = \sum_k p_k \phi(2x-k), \psi(x) = \sum_k g_k \phi(2x-k) \forall x \in R \qquad (4.2)$$

The functions $\phi(2x)$ and $\phi(2x-1)$ both are in V_1 and $V_1 = V_0 \oplus W_0$, thus there exist four sequences $\{a_{-2m}\}$, $\{b_{-2m}\}$, $\{a_{1-2m}\}$ and $\{b_{1-2m}\}$, $m \in Z$ such that

$$\phi(2x) = \sum_m [a_{-2m}\phi(x-m) + b_{-2m}\psi(x-m)],$$

$$\phi(2x-1) = \sum_m [a_{1-2m}\phi(x-m) + b_{1-2m}\psi(x-m)] \qquad , \qquad \forall x \in R \qquad .$$

$$(4.3)$$

From (4.3), we have

$$\phi(2x-l) = \sum_m [a_{l-2m}\phi(x-m) + b_{l-2m}\psi(x-m)] \qquad , \qquad l \in Z \qquad ,$$

$$(4.4)$$

which is defined as a decomposition relation of ϕ and ψ. We obtain two pairs of sequences $(\{p_k\}, \{q_k\})$ and $(\{a_k\}, \{b_k\})$ due to direct sum relationship, $V_1 = V_0 \oplus W_0$ these are unique. Here the sequences $\{p_k\}$ $\{q_k\}$ helped in reconstruction process and $\{a_k\}$ $\{b_k\}$ are used for the decomposition process. The series representation of all $f_j \in V_j$ and $h_j \in W_j$ is of the type :

$$f_j(x) = \sum_m c_{j,m}\phi(2^j x - m), c^j = \{c_{j,m}\} \in l^2. \qquad (4.5)$$

$$h_j(x) = \sum_m d_{j,m}\psi(2^j x - m), d^j = \{d_{j,m}\} \in l^2 \qquad (4.6)$$

Decomposition Algorithm Using (4.4), (4.5) and (4.6), we have

$$c_{j-1,k} = \sum_l a_{l-2k} \, c_{j,l} \qquad (4.7)$$

$$d_{j-1,k} = \sum_l b_{l-2k} \, c_{j,l} \qquad (4.8)$$

For a given scaling and wavelet function at any level the lower level coefficients are computed by recursively using the equations (4.7) and (4.8) respectively.

Reconstruction Algorithms Using (4.1), (4.5) and (4.6), we have

$$c_{j,k} = \sum_l \left[p_{k-2l} c_{j-1,l} + q_{k-2l} d_{j-1,l} \right]$$

By using the above equation at any level scaling coefficients are computed by using lower level coefficients.

5. Haar-Vilenkin Wavelets

The simplest example of wavelets is Haar wavelet and it generates classical multiresolution analysis, see for example [15]. The different generalization of Haar have been studied [5,9]. The extension of Haar function was studied by Vilenkin [21] and is often called the Haar-Vilenkin function. We have explained the concept of Haar-Vilenkin scaling and wavelet function in[17,18]. Haar-Vilenkin system is considered as follows:

Consider the set of positive integers which is denoted by P and suppose $k \in P$ is elaborated as

$$k = M_n + r(m_n - 1) + s - 1 \tag{5.1}$$

where $r = 0,1,...,M_n - 1$ and $s = 1,2,...,m_n - 1$ and n is from set of non-negative integers where N is the set of non-negative integers and $\{m_k\}_{k \in N}$ be a sequence of natural numbers where $m_k \geq 2$. Suppose for $k \in P$, $M_0 = 1$ and $M_k = m_{k-1} M_{k-1}$. Let $t \in [0,1)$ is arbitrary and it can be written as

$$t = \sum_{k=0}^{\infty} \frac{t_k}{M_{k+1}}, (0 \leq t_k < m_k) \tag{5.2}$$

Following function introduced by Vilenkin[14] and it is termed as the generalized Haar function[11]. Consider a sequence of functions $(h_k, k \in N)$ where $h_0 = 1$ and

$$h_k(t) = \begin{cases} \sqrt{M_n} \exp \dfrac{2\pi i s t_n}{m_n} & \dfrac{r}{M_n} \le t < \dfrac{r+1}{M_n} \\ 0 & otherwise \end{cases}$$

$$(5.3)$$

Also $h_k(t+1) = h_k(t) \forall\, t \in [0,1]$. On $L^2(R)$, the system of functions $\{h_k(t)\}$ is a complete orthonormal system. The function $h_k(t)$ for $k \in P$ and for $t \in [0,1)$ is a **mother wavelet** for where $h_k(t)$ is a **Haar-Vilenkin Wavelet** and the family $\{\psi_{a,b}(t)\}_{a,b \in Z}$ is Haar-Vilenkin system where $\psi_{a,b}(t) = m_n^{a/2} h_k(m_n^a t - b)$.

The system of functions is supported over $I_{a,b}$ where

$$I_{a,b} = \left[\frac{r}{m_n{}^a M_n} + \frac{b}{m_n{}^a}, \; \frac{r+1}{m_n{}^a M_n} + \frac{b}{m_n{}^a} \right[, \text{ for } a,b \in Z.$$

For $k \in P$ and for $t \in [0,1)$ as in equations (5.1) and (5.2) **the Haar-Vilenkin scaling function** is:

$$p_k(t) = \sqrt{M_n}\, \chi_{\left[\frac{r}{M_n}, \frac{r+1}{M_n} \right)}$$

$$= \begin{cases} \sqrt{M_n}, & \dfrac{r}{M_n} \le t < \dfrac{r+1}{M_n} \\ 0 & otherwise \end{cases}$$

$$(5.4)$$

Define $\phi_{a,b}(t) = m_n^{a/2} p_k(m_n^a t - b)$ and the system $\{\phi_{a,b}(t)\}_{a,b \in Z}$ is Haar-Vilenkin scaling functions system.

Remark 5.1: For $m_n=2$, Haar wavelet is a particular case of Haar-Vilenkin wavelet.

We have proved the various properties related to Haar-Vilenkin system like orthonormality of system, convergence of wavelet series and the properties of wavelet coefficients .

Effect of Jump Discontinuities on the behavior of Coefficients of Haar-Vilenkin System

Consider a function f(x) on the interval $\left[\frac{r}{M_n}, \frac{r+1}{M_n} \right]$ with jump discontinuity at $x_0 \in \left(\frac{r}{M_n}, \frac{r+1}{M_n} \right)$ and continuous at all other points in $\left[\frac{r}{M_n}, \frac{r+1}{M_n} \right]$. We have to find the value of $< f, \psi_{a,b} >$ i.e. Haar-Vilenkin coefficients for $x_0 \in I_{a,b}$ the coefficients for x_0 does not lies in $I_{a,b}$.

122

Suppose that on the intervals $\left[\frac{r}{M_n}, x_0\right]$ and $\left[x_0, \frac{r+1}{M_n}\right]$, the function f(x) is C². Consider $x_{a,b}$ as the mid point of $I_{a,b}$ for the fix integers $a \geq 0$ and $0 \leq b \leq m_n{}^a - 1$.

Case I: If x_0 *does not lies in* $I_{a,b}$, then we have

$$|< f, \psi_{a,b} >| \approx \frac{1}{4} m_n{}^{-3a/2} M_n{}^{-3/2}|f'(x_{a,b})|, \quad \text{for large values of a.}$$

Case II: If $x_0 \in I_{a,b}$, then if a is large, we have

$$|< f, \psi_{a,b} >| \approx m_n{}^{a/2} M_n{}^{1/2} \frac{1}{2m_n{}^a M_{n+1}} |f(x_0^-) - f(x_0^+)|$$

$$= \frac{m_n{}^{-a/2} M_n{}^{1/2}}{2M_{n+1}} |f(x_0^-) - f(x_0^+)|.$$

From the above two cases, we have the observation that for large values of a decay of $|< f, \psi_{a,b} >|$ is slower if $x_0 \in I_{a,b}$ rather than if x_0 *does not lies in* $I_{a,b}$.

5.1 A special type of Multiresolution Analysis

Definition 5.1 For k as in taken in (5.1), a special type of multi resolution analysis (MRA) is generated by a sequence which are closed subspaces $\{V_j\}_{j \in Z}$ of square integrable functions over R such that

1. $V_j \subset V_{j+1}$ for all integers j.

2. $\cup_{j \in Z} V_j$ is dense in L²(R).

3. $\cap_{j \in Z} V_j$ is a zero subspace.

4. $f(x) \in V_j$ iff $f(m_n{}^{-j}x) \in V_0$ for all integers j.

5. There exists $g_k(x)$ in L²(R) , such that the system of translates of g_k i.e. $\left\{g_k\left(t - \frac{b}{M_n}\right)\right\}_{b \in Z}$ is an orthonormal system, here $g_k(x)$ is called a scaling function and

$$V_0 = \overline{span}\left\{T_{\frac{b}{M_n}} g_k(x)\right\}.$$

Remark 5.2 In order to define a special type of MRA, first we identify the space V_0 and then take V_j as

$$V_j = \{f(x): f(x) = D_{m_n{}^j} g_k(x), g_k(x) \in V_0\}$$

so that the Definition 5.1(4) is satisfied and then we prove that the conditions (1), (2),(3) and (5) of Definition 5.1 hold. First identify the function $g_k(x)$ and define V_0 such that the system of translates i.e. $\{T_{\frac{b}{M_n}} g_k(x)\}_{b \in Z}$ over Z is an orthonormal system and take

$$V_0 = \overline{span}\left\{T_{\frac{b}{M_n}} g_k(x)\right\}.$$

Example 5.1 [Haar-Vilenkin MRA] Consider the set V_0 of the function $f(x)$ which are the step functions and satisfy

    (i) $f(x)$ is square integrable over R.

    (ii)   Over the interval $I_{0,\frac{b}{M_n}} = \left[\frac{r+b}{M_n}, \frac{r+b+1}{M_n}\right[$, f is constant $\forall b \in Z$.

It can be verified that for $l \in Z$, $V_0 = \overline{span}\left\{T_{\frac{l}{M_n}} g_k(x)\right\}$ where

$$p_k(x) = \sqrt{M_n}\, \chi_{\left[\frac{r}{M_n}, \frac{r+1}{M_n}\right)}.$$

5.2 Integral forms of Haar-Vilenkin Wavelets

Define the Haar-Vilenkin system $(h_k, k \in N)$ taken as over [A,B] of length 1 as $h_0 = 1$ and

$$h_k(t) = \begin{cases} \sqrt{M_n} \exp\dfrac{2\pi i s t_n}{m_n} & A + \dfrac{r}{M_n} \le t < A + \dfrac{r+1}{M_n} \\ 0 & otherwise \end{cases}$$

$$(5.5)$$

for k as defined in 5.1, see[19]. Let

$$P_{v,i}(x) = \int_A^x \int_A^x \cdots \int_A^x h_i(t)dt^v$$

$$= \frac{1}{(v-1)!} \int_A^x (x-t)^{v-1} h_i(t)dt.$$

For $i \neq 1$ and for $\dfrac{r}{M_n} \le x < \dfrac{r}{M_n} + \dfrac{1}{M_{n-1}}$, we have

$$P_{\alpha,i}(x) = \frac{1}{(\alpha-1)!} \int x(x-t)^{\alpha-1} \sqrt{M_n}\, dt$$

$$= \frac{\sqrt{M_n}}{(\alpha-1)!} \int x(x-t)^{\alpha-1}\, dt$$

$$= \frac{\sqrt{M_n}}{(\alpha-1)!} \frac{1}{\alpha}\left(x - \frac{r}{M_n}\right)^\alpha$$

$$= \frac{\sqrt{M_n}}{\alpha!}\left(x - \frac{r}{M_n}\right)^\alpha$$

124

If $\frac{r}{M_n} + \frac{1}{M_{n-1}} \leq x < \frac{r}{M_n} + \frac{2}{M_{n-1}}$, then on solving as above, we have

$$P_{\alpha,i}(x) = \frac{\sqrt{M_n}}{\alpha!}\left[x - \left(\frac{r}{M_n} + \frac{1}{M_{n-1}}\right)\right]^\alpha e^{2\pi i s/m_n}$$

...

...

If $\frac{r}{M_n} + \frac{m_n-1}{M_{n-1}} \leq x < \frac{r+1}{M_n}$, we have

$$P_{\alpha,i}(x) = \frac{\sqrt{M_n}}{\alpha!}\left[x - \left(\frac{r}{M_n} + \frac{m_n-1}{M_{n-1}}\right)\right]^\alpha e^{2\pi i s(m_n-1)/m_n}.$$

Thus
$P_{\alpha,i}(x) =$

$$\begin{cases} 0 & x < \frac{r}{M_n} \\[2ex] \frac{\sqrt{M_n}}{\alpha!}\left(x - \frac{r}{M_n}\right)^\alpha & \frac{r}{M_n} \leq x < \frac{r}{M_n} + \frac{1}{M_{n-1}} \\[2ex] \frac{\sqrt{M_n}}{\alpha!}\left[x - \left(\frac{r}{M_n} + \frac{1}{M_{n-1}}\right)\right]^\alpha e^{2\pi i s/m_n} & \frac{r}{M_n} + \frac{1}{M_{n-1}} \leq x < \frac{r}{M_n} + \frac{2}{M_{n-1}} \\[1ex] \qquad\qquad .... & \qquad ... \\[1ex] \frac{\sqrt{M_n}}{\alpha!}\left[x - \left(\frac{r}{M_n} + \frac{m_n-1}{M_{n-1}}\right)\right]^\alpha e^{2\pi i s(m_n-1)/m_n} & \frac{r}{M_n} + \frac{m_n-1}{M_{n-1}} \leq x < \frac{r+1}{M_n} \end{cases}$$

$$(5.6)$$

We have $h_i(t) = 0$ for $i=0$ and

$$P_{\alpha,1}(x) = \frac{1}{(\alpha-1)!}\int_A^x (x-t)^{\alpha-1}\, dt.$$

$$= \frac{1}{(\alpha-1)!}\frac{1}{\alpha}(x-A)^\alpha$$

$$= \frac{(x-A)^\alpha}{(\alpha)!}$$

$$(5.7)$$

Thus we have equation (5.6) for $i > 1$ and (5.7) for i=1.

5.3 Haar-Vilenkin Wavelets in Matrix Form

We have constructed the wavelets in discrete form for the case where A=0 and B=1. The grid points are denoted as

$$\tilde{x}_l = A + l\,\delta x, \quad l = 0,1,2,\dots,m_0. \quad (5.8)$$

We have considered

$$x_l = \frac{1}{2}(\widetilde{x_{l-1}} + \widetilde{x_l}), l = 1,2, \dots, m_0$$

$$(5.9)$$

We get the Haar-Vilenkin wavelets on replacing x by x_l in equations (5.5), (5.6) and (5.7). The elements of square matrices H, P_1, P_2, ..., P_v, that we have introduced are

$$H(i,l) = h_i(x_l), P_v(i,l) = P_{v_i}(x_l), v = 1,2,3 \dots$$

A=0, B=1 and

$$\delta x = \frac{B - A}{m_0} = \frac{1}{m_0}.$$

Example 5.2 Take a sequence $(m_k, k \in N) = (2,2,2, \dots)$.
Then $x_1=1/4$ and $x_3=3/4$. Then

$$h_1(t) = \begin{cases} 1 & 0 \le t < 1/2 \\ -1 & \frac{1}{2} \le t < 1 \\ 0 & otherwise \end{cases}$$

$$h_1(t) = \begin{cases} \sqrt{2} & 0 \le t < 1/4 \\ -\sqrt{2} & \frac{1}{4} \le t < 1/2 \\ 0 & otherwise \end{cases}$$

The Haar-Vilenkin matrix is formulated as

$$H = \begin{bmatrix} h_1(x_1) & h_1(x_2) \\ h_2(x_1) & h_2(x_2) \end{bmatrix} = \begin{bmatrix} 1 & -1 \\ \sqrt{2} & 0 \end{bmatrix}.$$

The other matrices are

$$P_1 = \begin{bmatrix} P_{11}(x_1) & P_{11}(x_2) \\ P_{12}(x_1) & P_{12}(x_2) \end{bmatrix}, P_2 = \begin{bmatrix} P_{21}(x_1) & P_{21}(x_2) \\ P_{22}(x_1) & P_{22}(x_2) \end{bmatrix} \dots$$

Using the equations (5.6) and (5.7), we have

$$P_1 = \begin{bmatrix} \frac{1}{4} & \frac{3}{4} \\ \frac{1}{32} & \frac{9}{32} \end{bmatrix}.$$

6. Conclusion

We have explained the basic properties of wavelets, multiresolution analysis, the process of construction and decomposition of wavelets in this chapter. We have also studied Haar-Vilenkin wavelets which are also termed as generalized Haar wavelets. The Haar-Vilenkin wavelets have also been studied in integral and matrix form. These methods will be useful in solution of ordinary and partial differential equations.

References

1. Aldroubi, A. and Unser, M.A. (eds) (1996). Wavelets in Medicine and Biology. CRC Press, Boca Raton, FL.
2. Benedetto, J.J. and Frazier, M. (eds) (1994). Wavelets: Mathematics and Applications. CRC Press, Boca Raton, FL.
3. Chambolle, A., Devore, R.A., Lee, N.Y. and Lucier, B.J. (1998). Nonlinear wavelet image processing: Variational problems, compression and noise removal through wavelet shrinkage. IEEE Transactions on Image Processing, 7, 319-335.
4. Ciesielski, Z. (1985). Haar orthogonal Functions in Analysis and Probability. Colloquia Societatis James Bolyai, 49, Alfred Haar Memorial Conference, Budapest, 25-57.
5. Christensen, O. and Christensen, K.L. (2004). Approximation Theory- From Taylor Polynomials to Wavelets. Birkhauser, Boston.
6. Chui, C.K. (1992). An Introduction to Wavelets. Academic Press, San Diego.
7. Cohen, A. (2003). Numerical Analysis of Wavelet methods. Elsevier, Amsterdam.
8. Cohen, A. and Ryan, R.D. (1995). Wavelets and multiscale signal processing. Applied Mathematics and Mathematical Computation 11
9. Coifman, R.R, Meyer, Y., Quake, S.R. and Wickerhauser, M.V. (1992). Signal processing and compression with wavelet packets. In: Meyer, Y. and Roques, S.(eds). Progress in Wavelet Analysis and Applications. pp. 77-93. Editions Frontieres.
10. Daubechies, I. (1992). Ten Lectures on Wavelets, Society of Industrial and Applied Mathematics, Philadelphia
11. Gabor, D. (1946). Theory of communication. Journal of IEE 93: 429-457.
12. Georgiou, E.F. and Kumar, P. (eds)(1994). Wavelets in Geophysics. Academic Press, San Diego.
13. Grossman, A. and Morlet, J. (1984). Decomposition of Hardy functions into square-integrable wavelets of constant shapes. SIAM *Journal on Mathematical Analysis* **15(4):** 723-736.
14. Haar, A. (1910). Zur theorie der orthogonalen funktionen-systeme. Mathematische An-nalen 69: 331-371.
15. Kobayashi, M. (ed) (1998). Wavelets and their Applications: Case Studies. SIAM, Philadelphia.

16. Mallat, S. (1999). A Wavelet Tour of Signal Processing. 2nd Edition, Academic Press, San Diego.
17. Manchanda, P. and Meenakshi (2009). New classes of wavelets. In: Siddiqi, A.H., Gupta, A.K. and Brokate, M. (eds). Modelling of Engineering and Technological Problems. pp253-271. AIP Conference proceedings 1146, AIP, NewYork.
18. Manchanda, P., Meenakshi and Siddiqi, A.H. (2008). Haar-Vilenkin wavelet. The Aligarh Buletin of Mathematics 27(1): 59-73.
19. Meenakshi, Manchanda P. (2017). Construction and Properties of Haar-Vilenkin Wavelets.In: Industrial Mathematics and Complex Systems.
20. Meyer, Y. (1992). Wavelets and Operators. Cambridge University Press, Cambridge, UK
21. F. Schipp, W. R. Wade and P. Simon (1990). Walsh series: An Introduction to Dyadic Harmonic Analysis, Adam Hilger Ltd., Bristol and New York.
22. Walnut, D. (2001). An Introduction to Wavelet Analysis. Birkhauser, Boston.
23. Woztaszczyk, P. (1997). A Mathematical Introduction to Wavelets. London Mathematical Society Student Texts 37, Cambridge University Press, Cambridge, UK.

CHAPTER 8

A Mathematical Study of Phytoplankton and Zooplankton Model with Time Delay

Rakesh Kumar[1], Navneet Rana[2] and Anuj Kumar Sharma[3]
[1]Shaheed Bhagat Singh State University, Ferozepur, Punjab, India
[2]Guru Nanak College, Sri Muktsar Sahib, Punjab, India
[3]Lala Lajpat Rai DAV College, Jagraon, Punjab, India

1. Introduction

Plankton are very large collection of microorganisms present in water. In water, they are source of food to many small and large aquatic organisms. Plankton are necessary element for existence of living beings on earth. Plankton play an integral part to support the health and stability between the ocean and its complicated food chains. The life of living beings on earth is greatly supported by the oxygen, nutrients, and biomass produced by these aquatic plants. These plants contribute about 70% of the oxygen in atmosphere. Such types of plants present in water are known as phytoplankton. The phytoplankton is responsible to absorb carbon dioxide and produce oxygen with the help of process of photosynthesis. Prochlorococcus is a type of phytoplankton which produces maximum oxygen in the atmosphere. Zooplanktons are heterotrophic plankton, which depend upon the phytoplankton for their food and thus gain energy from them. When these organisms are eaten by some large predators, then this energy is transferred to them. We can say that zooplankton act as bridge between phytoplankton and large predators such as fish. In aquatic ecology, plankton are prime part and their growth is greatly affected by the temperature, salinity, pH level and nutrient level present in water. Many mathematical models are suggested by researchers to study how these changes effect the plankton population in water. The authors in [1-4] studied the interaction of phytoplankton and zooplankton population with toxicity. In [5], the authors explained the effect of water temperature on growth of plankton. The authors in [6-11] modified the model by considering time delay parameter and showed that stability of the system depends upon the delay parameter. In [12], the authors described the phytoplankton-zooplankton system under the influence of noise. The effect of fear on prey predator system is explained by authors in [13]. The authors in [14] outlined the bifurcation analysis of a nonlinear model.

The present paper is arranged in different sections as: The mathematical settings to formulate the problem are done in second section. In third section, the positivity and boundedness of solution has been discussed. The equilibrium points, their stability analysis and the state for presence of Hopf bifurcation is derived in fourth section. Further in fifth section, numerical simulations are completed to verify the conceptual results. At last, the concluding remark is given.

2. Mathematical description of delay model

Let $q_1(t)$ and $q_2(t)$ be the density of phytoplankton population and zooplankton population respectively. Let τ be the time delay in zooplankton predation, arises when zooplankton population spent time in journey which occur either in vertical direction or in horizontal direction produced by the higher predators like fish. The basic model can be represented mathematically by the set of ordinary differential equations as:

$$\frac{dq_1(t)}{dt} = aq_1\left(1 - \frac{q_1}{K}\right) - \frac{\alpha q_1(t)q_2(t-\tau)}{\beta + q_1(t)} - \delta_1 q_1(t),$$

$$\frac{dq_2(t)}{dt} = \frac{\alpha_1 \alpha q_1(t)q_2(t-\tau)}{\beta + q_1(t)} - \delta_2 q_2(t) - \frac{\theta q_1^2 q_2}{\gamma^2 + q_1^2}.$$

(1)

The system (1) is defined with initial conditions as:

$q_1(\phi) = \psi_1(\phi), q_2(\phi) = \psi_2(\phi); \ \psi_1(\phi) \geq 0, \psi_2(\phi) \geq 0, \phi \in [-\tau, 0]$ and $\psi_1(0) > 0, \psi_2(0) > 0$.

Also, $\psi_1(\phi), \psi_2(\phi) \in C([-\tau, 0], R_+^2)$ where $C([-\tau, 0], R_+^2)$ is the Banach space of continuously defined functions on the interval $[-\tau, 0]$ into $R_+^2 = \{(x_1, x_2) : x_i > 0, i = 1, 2\}$.

The various parameters involved in system (1) are described in Table (1):

Table 1: Explanation of parameters

Parameters	Explanation
a	Rate at which phytoplankton grows naturally
K	Phytoplankton'scarrying capacity
α	Predation rate of zooplankton over phytoplankton
α_1	Conversion efficiency of phytoplankton into zoo-plankton
β	Half saturation constant
δ_1	Death rate of phytoplankton
δ_2	Death rate of zooplankton
$\dfrac{\theta q_1^{\ 2}}{\gamma^2 + q_1^{\ 2}}$	Holling type III functional response

3. Positivity and boundedness of solution

As we are dealing with system of biological existence, it is important to check the system for positive and bounded solution. In this section, we are going to analyse the system (1) for solution which is positive and bounded.

Theorem 1: The solution of the system (1) is positive corresponding to initial conditions $\psi_1(\phi) > 0, \psi_2(\phi) > 0$ defined on $[0, +\infty)$.

Proof: Let $\left(q_1(t), q_2(t)\right)$ be solution of the system (1) corresponding to the initial conditions. We can rewrite the first equation of the system (1) as

$$\frac{dq_1}{q_1} = \left[a\left(1 - \frac{q_1}{K}\right) - \frac{\alpha q_2(t - \tau)}{\beta + q_1(t)} - \delta_1 \right] dt,$$

$$\frac{dq_1}{q_1} = G_1 dt, \qquad (2)$$

where

$$G_1 = a\left(1 - \frac{q_1}{K}\right) - \frac{\alpha q_2(t - \tau)}{\beta + q_1(t)} - \delta_1,$$

Integrate the equation (2) with limits 0 to t, we get the result as:

$$q_1(t) = \psi_1(0) \exp(G_{11}),$$

where $G_{11} = \int_0^t G_1 dt$.

We can rewrite the second equation of system (1) as:

$$\frac{dq_2}{dt} \geq -\delta_2 q_2(t) - \frac{\theta q_1^2 q_2}{\gamma^2 + q_1^2},$$

$$\frac{dq_2}{q_2} \geq \left[-\delta_2 - \frac{\theta q_1^2}{\gamma^2 + q_1^2} \right] dt,$$

$$\frac{dq_2}{q_2} \geq G_2 dt, \tag{3}$$

Where $G_2 = -\delta_2 - \frac{\theta q_1^2}{\gamma^2 + q_1^2}$,

Integrate the equation (3) with limits 0 to t, we get the result as:

$$q_2(t) \geq \psi_2(0) \exp(G_{22}),$$

Where $G_{22} = \int_0^t G_2 dt$.

Hence, $q_1(t) > 0, q_2(t) > 0$ i.e. the solution with positive initial conditions is positive.
In biological system, no population becomes indefinitely large. So, it is very essential to check the boundedness of solution.

Theorem 2: All the solutions of the system of ordinary differential equation involved in model (1) which initiates in R_+^2 are consistently bounded.
Proof: Let $(q_1(t), q_2(t))$ be solution of the system (1) w.r.t. the positive initial conditions. The first equation of the system (1) can be reported as:

$$\frac{dq_1(t)}{dt} \leq aq_1 \left(1 - \frac{q_1}{K} \right),$$

By the result of comparison theorem, we get that $\lim_{t \to \infty} \sup q_1(t) \leq K$.

Consider a time dependent function $\Lambda(t) = \alpha_1 q_1 + q_2$, then

$$\frac{d\Lambda}{dt} = \alpha_1 a q_1 - \frac{\alpha_1 a q_1^2}{K} - \alpha_1 \delta_1 q_1 - \delta_2 q_2 - \frac{\theta q_1^2 q_2}{\gamma^2 + q_1^2},$$

$$\frac{d\Lambda}{dt} \leq \alpha_1 a q_1 - \alpha_1 \delta_1 q_1 - \delta_2 q_2,$$

$$\frac{d\Lambda}{dt} + \delta_2 \Lambda \leq \alpha_1 \left(a - \delta_1 + \delta_2 \right) q_1,$$

By simpler calculations [15], we obtain

$$\lim_{t \to \infty} \sup \Lambda(t) = \lim_{t \to \infty} \sup \left(\alpha_1 q_1(t) + q_2(t) \right) \leq \frac{\alpha_1 \left(a - \delta_1 + \delta_2 \right)}{\delta_2} K.$$

Therefore, all the solutions of system of differential equations involved in model (1) are consistently bounded.

4. Equilibrium points and their stability analysis

Model (1) has three feasible steady state points, which are defined as:

1. The trivial equilibrium point $E_0(0,0)$ always exists.

2. The zooplankton free equilibrium point $E_1(q_1,0)$, where

   $$q_1 = \frac{K(a - \delta_1)}{a}$$ which occurs when there is no zooplankton

   population appear in the system.

3. The positive equilibrium point $E_2(q_1^*, q_2^*)$ is the solution of the set of equations

   $$aq_1^* \left(1 - \frac{q_1^*}{K} \right) - \frac{\alpha q_1^* q_2^*}{\beta + q_1^*} - \delta_1 q_1^* = 0,$$

   $$\frac{\alpha_1 \alpha q_1^* q_2^*}{\beta + q_1^*} - \delta_2 q_2^* - \frac{\theta q_1^{*2} q_2^*}{\gamma^2 + q_1^{*2}} = 0.$$

On solving this system of equations, we get q_1^* is root of equation $a_3 q_1^{*3} + a_2 q_1^{*2} + a_1 q_1^* + a_0 = 0$, $\quad$ (4)

Where

$$a_3 = \alpha_1 \alpha - \delta_2 - \theta, a_2 = -\beta(\delta_2 + \theta), a_1 = \gamma^2 (\alpha_1 \alpha - \delta_2), a_0 = -\beta \delta_2 \gamma^2$$

and $q_2^* = \frac{\beta + q_1^*}{\alpha} \left(a - \frac{a q_1^*}{K} - \delta_1 \right)$.

133

As the coefficients of cubic equation satisfy the condition $a_2 < 0$ and $a_0 < 0$. If $\alpha_1 \alpha > \delta_2$, then $a_1 > 0$. Then the equation (4) has unique positive root if $\Delta < 0$ and the equation (3) has three distinct   positive   root   if $\Delta > 0$,

where $\Delta = 18 a_3 a_2 a_1 a_0 - 4 a_2^3 a_0 + a_2^2 a_1^2 - 4 a_3 a_1^3 - 27 a_3^2 a_0^2$.

We use the Jacobian matrix in order to study the behaviour of solution near all the steady state points. The Jacobian matrix calculated at any point (q_1, q_2) is given by

$$J_E = \begin{pmatrix} a_{11} & a_{12} \\ a_{21} & a_{22} \end{pmatrix}$$

Where

$$a_{11} = a - \frac{2 a q_1}{K} - \frac{\alpha \beta q_2 (t - \tau)}{(\beta + q_1)^2} - \delta_1;$$

$$a_{12} = -\frac{\alpha q_1 e^{-\lambda \tau}}{\beta + q_1};$$

$$a_{21} = \frac{\alpha_1 \alpha \beta q_2 (t - \tau)}{(\beta + q_1)^2} - \frac{2 \theta \gamma^2 q_1 q_2}{(\gamma^2 + q_1^2)^2};$$

$$a_{22} = \frac{\alpha_1 \alpha q_1 e^{-\lambda \tau}}{\beta + q_1} - \delta_2 - \frac{\theta q_1^2}{\gamma^2 + q_1^2}.$$

4.1 Stability analysis

In present part, we analyse the stability behaviour of the system around the equilibrium points. For analysis of local stability around equilibrium point, we find the condition for which Jacobian matrix has eigen-values with negative real part.

Theorem 3: The trivial equilibrium point $E_0(0,0)$ is reliable if $a < \delta_1$.

Proof: The equilibrium point $E_0(0,0)$ has Jacobian matrix, defined by

$$J_{E_0} = \begin{pmatrix} a - \delta_1 & 0 \\ 0 & -\delta_2 \end{pmatrix}$$

The latent roots of J_{E_0} are $a - \delta_1, -\delta_2$. Therefore, the trivial state steady point $E_0(0,0)$ is stable if $a < \delta_1$.

Theorem 4: The zooplankton free equilibrium point $E_1(q_1, 0)$ without any time delay, where $q_1 = \dfrac{K(a - \delta_1)}{a}$ is LSA

if $a > \delta_1$ and $\dfrac{\alpha_1 \alpha K(a - \delta_1)}{\beta a + K(a - \delta_1)} < \delta_2 + \dfrac{\theta K^2(a - \delta_1)^2}{\gamma^2 a^2 + K^2(a - \delta_1)^2}$.

Proof: The equilibrium point $E_1(q_1, 0)$ has Jacobian matrix defined by

$$J_{E_1} = \begin{pmatrix} a - \dfrac{2aq_1}{K} - \delta_1 & \dfrac{-\alpha q_1}{\beta + q_1} \\ 0 & \dfrac{\alpha_1 \alpha q_1}{\beta + q_1} - \delta_2 - \dfrac{\theta q_1^2}{\gamma^2 + q_1^2} \end{pmatrix}$$

By using the value of $q_1 = \dfrac{K(a - \delta_1)}{a}$, the eigenvalues of J_{E_1}

are $-a + \delta_1$ and $\dfrac{\alpha_1 \alpha K(a - \delta_1)}{\beta a + K(a - \delta_1)} - \delta_2 - \dfrac{\theta K^2(a - \delta_1)^2}{\gamma^2 a^2 + K^2(a - \delta_1)^2}$. These

eigenvalues are negative provided $a > \delta_1$ and

$\dfrac{\alpha_1 \alpha K(a - \delta_1)}{\beta a + K(a - \delta_1)} < \delta_2 + \dfrac{\theta K^2(a - \delta_1)^2}{\gamma^2 a^2 + K^2(a - \delta_1)^2}$.

Hence, zooplankton free steady state $E_1(q_1, 0)$, where

$q_1 = \dfrac{K(a - \delta_1)}{a}$ without any time delay is LAS if

$a > \delta_1$ and $\dfrac{\alpha_1 \alpha K(a - \delta_1)}{\beta a + K(a - \delta_1)} < \delta_2 + \dfrac{\theta K^2(a - \delta_1)^2}{\gamma^2 a^2 + K^2(a - \delta_1)^2}$.

Next, we are presenting the stability conditions of positive equilibrium point $E_2(q_1^*, q_2^*)$ show that stability disappears through Hopf bifurcation and as a result of which instability takes place. The bifurcation parameter is time delay (τ). First, we linearize the system

135

(1) around E_2 to get the required characteristic equation. The Jacobian matrix at equilibrium point $E_2(q_1^*, q_2^*)$ is given by

$$J_{E_2} = \begin{pmatrix} a - \dfrac{2aq_1^*}{K} - \dfrac{\alpha\beta q_2^*(t-\tau)}{\left(\beta+q_1^*\right)^2} - \delta_1 & \dfrac{-\alpha q_1^* e^{-\lambda\tau}}{\beta+q_1^*} \\[3mm] \dfrac{\alpha_1\alpha\beta q_2^*(t-\tau)}{\left(\beta+q_1^*\right)^2} - \dfrac{2\theta\gamma^2 q_1^* q_2^*}{\left(\gamma^2+q_1^{*2}\right)^2} & \dfrac{\alpha_1\alpha q_1^* e^{-\lambda\tau}}{\beta+q_1^*} - \delta_2 - \dfrac{\theta q_1^{*2}}{\gamma^2+q_1^{*2}} \end{pmatrix}$$

The Jacobian matrix J_{E_2} has characteristic equation defined as

$$\Delta(\lambda,\tau)\big|_{E_2} = \lambda^2 + Y_1\lambda + Y_2 + (Y_3\lambda + Y_4)e^{-\lambda\tau} = 0 \tag{5}$$

Where

$$Y_1 = -a + \frac{2aq_1^*}{K} + \frac{\alpha\beta q_2^*(t-\tau)}{(\beta+q_1^*)^2} + \delta_1 + \delta_2 + \frac{\theta q_1^{*2}}{\gamma^2+q_1^{*2}};$$

$$Y_2 = -a\delta_2 - a\frac{\theta q_1^{*2}}{\gamma^2+q_1^{*2}} + \frac{2aq_1^*\delta_2}{K} + \frac{2a\theta q_1^{*3}}{K\left(\gamma^2+q_1^{*2}\right)} + \frac{\alpha\beta\delta_2 q_2^*(t-\tau)}{\left(\beta+q_1^*\right)^2} + \frac{\alpha\beta q_2^*(t-\tau)}{\left(\beta+q_1^*\right)^2}\frac{\theta q_1^{*2}}{\left(\gamma^2+q_1^{*2}\right)} + \delta_1\delta_2 + \delta_1\frac{\theta q_1^{*2}}{\gamma^2+q_1^{*2}};$$

$$Y_3 = -\frac{\alpha_1\alpha q_1^*}{\beta+q_1^*};$$

$$Y_4 = a\frac{\alpha_1\alpha q_1^*}{\beta+q_1^*} - 2a\frac{\alpha_1\alpha q_1^{*2}}{K\left(\beta+q_1^*\right)} - \frac{\alpha_1\alpha q_1^*\delta_1}{\beta+q_1^*} - \frac{2\alpha q_1^{*2}\theta\gamma^2 q_2^*}{\left(\beta+q_1^*\right)\left(\gamma^2+q_1^{*2}\right)^2}.$$

The values of Y_1, Y_2, Y_3, Y_4 are dependent on τ. Also, Y_1, Y_2, Y_3, Y_4 are all continuous and differentiable functions.

Theorem 4: The positive equilibrium point $E_2(q_1^*, q_2^*)$ with no time delay is asymptotically stable

if $\dfrac{2aq_1^*}{K} + \dfrac{\alpha\beta q_2^*(t-\tau)}{(\beta+q_1^*)^2} + \delta_1 + \delta_2 + \dfrac{\theta q_1^{*2}}{\gamma^2+q_1^{*2}} > a + \dfrac{\alpha_1\alpha q_1^*}{\beta+q_1^*}$ and

$$\frac{2aq_1^*\delta_2}{K} + \frac{2a\theta q_1^{*3}}{K\left(\gamma^2+q_1^{*2}\right)} + \frac{\alpha\beta\delta_2 q_2^*(t-\tau)}{\left(\beta+q_1^*\right)^2} + \frac{\alpha\beta q_2^*(t-\tau)}{\left(\beta+q_1^*\right)^2}\frac{\theta q_1^{*2}}{\left(\gamma^2+q_1^{*2}\right)} + \delta_1\delta_2 + \delta_1\frac{\theta q_1^{*2}}{\gamma^2+q_1^{*2}} + a\frac{\alpha_1\alpha q_1^*}{\beta+q_1^*} >$$

$$a\delta_2 + a\frac{\theta q_1^{*2}}{\gamma^2+q_1^{*2}} + 2a\frac{\alpha_1\alpha q_1^{*2}}{K\left(\beta+q_1^*\right)} + \frac{\alpha_1\alpha q_1^*\delta_1}{\beta+q_1^*} + \frac{2\alpha q_1^{*2}\theta\gamma^2 q_2^*}{\left(\beta+q_1^*\right)\left(\gamma^2+q_1^{*2}\right)^2}$$

Proof: For $\tau = 0$, the equation (5) becomes

$$\Delta(\lambda,0)\big|_{E_2} = \lambda^2 + (Y_1+Y_3)\lambda + (Y_2+Y_4) = 0 \tag{6}$$

By Routh-Hurwitz criterion [16], the system represented by equation (1) is asymptotically stable if the characteristics equation has all latent roots with negative real parts i.e. if

$$(A_1): Y_1 + Y_3 > 0 \text{ and } (A_2): Y_2 + Y_4 > 0$$

which gives

$$Y_1 + Y_3 = -a + \frac{2aq_1^*}{K} + \frac{\alpha\beta q_2^*(t-\tau)}{(\beta+q_1^*)^2} + \delta_1 + \delta_2 + \frac{\theta q_1^{*2}}{\gamma^2 + q_1^{*2}} - \frac{\alpha_1 \alpha q_1^*}{\beta + q_1^*} > 0$$

$$\Rightarrow \frac{2aq_1^*}{K} + \frac{\alpha\beta q_2^*(t-\tau)}{(\beta+q_1^*)^2} + \delta_1 + \delta_2 + \frac{\theta q_1^{*2}}{\gamma^2 + q_1^{*2}} > a + \frac{\alpha_1 \alpha q_1^*}{\beta + q_1^*}$$

and

$$Y_2 + Y_4 = -a\delta_2 - a\frac{\theta q_1^{*2}}{\gamma^2 + q_1^{*2}} + \frac{2aq_1^* \delta_2}{K} + \frac{2a\theta q_1^{*3}}{K(\gamma^2 + q_1^{*2})} + \frac{\alpha\beta\delta_2 q_2^*(t-\tau)}{(\beta+q_1^*)^2} + \frac{\alpha\beta q_2^*(t-\tau)}{(\beta+q_1^*)^2} \frac{\theta q_1^{*2}}{(\gamma^2 + q_1^{*2})}$$

$$+\delta_1\delta_2 + \delta_1 \frac{\theta q_1^{*2}}{\gamma^2 + q_1^{*2}} + a\frac{\alpha_1 \alpha q_1^*}{\beta + q_1^*} - 2a\frac{\alpha_1 \alpha q_1^{*2}}{K(\beta+q_1^*)} - \frac{\alpha_1 \alpha q_1^* \delta_1}{\beta + q_1^*} - \frac{2aq_1^{*2} \theta \gamma^2 q_2^*}{(\beta+q_1^*)(\gamma^2 + q_1^{*2})^2} > 0$$

$$\Rightarrow \frac{2aq_1^* \delta_2}{K} + \frac{2a\theta q_1^{*3}}{K(\gamma^2 + q_1^{*2})} + \frac{\alpha\beta\delta_2 q_2^*(t-\tau)}{(\beta+q_1^*)^2} + \frac{\alpha\beta q_2^*(t-\tau)}{(\beta+q_1^*)^2} \frac{\theta q_1^{*2}}{(\gamma^2 + q_1^{*2})} + \delta_1\delta_2 + \delta_1 \frac{\theta q_1^{*2}}{\gamma^2 + q_1^{*2}} + a\frac{\alpha_1 \alpha q_1^*}{\beta + q_1^*} >$$

$$a\delta_2 + a\frac{\theta q_1^{*2}}{\gamma^2 + q_1^{*2}} + 2a\frac{\alpha_1 \alpha q_1^{*2}}{K(\beta+q_1^*)} + \frac{\alpha_1 \alpha q_1^* \delta_1}{\beta + q_1^*} + \frac{2aq_1^{*2} \theta \gamma^2 q_2^*}{(\beta+q_1^*)(\gamma^2 + q_1^{*2})^2}.$$

4.2 Analysis of model with time delay

Here we consider positive time delay i.e. $\tau > 0$. Now, we focus to check the stability conditions of system (1) about $E_2(q_1^*, q_2^*)$. For this, we calculate the roots of equation (5) and inspect that with the change in value of τ, the real part of root rises to become zero and afterwards a positive number. This helps us to show that time delay parameter causes Hopf bifurcation to take place.

Remark: [5, 9]

If (A_1) and (A_2) hold, then for $\tau = 0$, the equation (5) have negative real part of root. It means that positive steady state point is locally stable. The equation (5) has purely imaginary roots if real part of root is zero for some $\tau > 0$. This is possible by the condition of continuity of solutions and roots with nonnegative real part of characteristic equation (5).

By the result of [9], the positive steady point changes to unstable depending upon the two conditions: For some $\tau > 0$, if

$\Delta(0,\tau)=0$ or pair of complex roots which are conjugate to each other exist for characteristic equation (5). The value of $\Delta(0,\tau)=Y_2+Y_4\neq 0$ and thus the first condition does not hold. It means that the positive steady point changes into unstable due to the second condition defined as the characteristic equation (5) has a conjugate pair of complex values and these complex values occurs at critical values of $\tau_0^* > 0$.

Thus, τ_0^* is the least possible value at which a complex values exists pairwise which is conjugate to each other and at $\tau=\tau_0^*$, all the eigenvalues which are left must have negative real parts. This gives us the various cases for the positive steady state point of system (1) to be absolute and conditional stable.

The case for non-occurrence of time delayed instability, as mentioned in Gopalsamy [17] is stated as:

Theorem 5: The necessary and sufficient conditions for positive steady state point in the presence of time delay to be LAS are:
  (a) The characteristic equation (5) has all values with negative real parts.
  (b) For $\tau>0$ and all real ξ, $\Delta(\lambda,\tau)\neq 0$.

Next, we are presenting the condition of occurrence of Hopf bifurcation of positive steady point $E_2(q_1^*,q_2^*)$ in the time delay model represented in equation (1). The time delay represented by τ is taken as the control parameter of bifurcation.

Theorem 6: If (A_3) holds,

where $(A_3):Y_1^2-Y_3^2-2Y_2>0, Y_2^2-Y_4^2>0$ then the positive steady

point $E_2(q_1^*,q_2^*)$ is LAS for time delay $\tau>0$.

Proof: Here, we are going to calculate complex values of characteristic equation (5) that occur pairwise.

Let, for a $\tau>0$, the root of characteristic equation (5) is taken as $\lambda=i\xi$ ($\xi>0$ and $i=\sqrt{-1}$), ξ is positive real. Put $\lambda=i\xi$ in (5) and using the methods mentioned in [18, 19] for separating different parts of characteristic equation (5), we get the real and imaginary parts as

$$Y_4 \cos \xi \tau + Y_3 \xi \sin \xi \tau = \xi^2 - Y_2 , \tag{7}$$

$$Y_3 \xi \cos \xi \tau - Y_4 \sin \xi \tau = -Y_1 \xi . \tag{8}$$

By using simple methods of squaring and adding equation (7) and (8), we get a 4th order equation in ξ as

$$\xi^4 + (Y_1^2 - Y_3^2 - 2Y_2)\xi^2 + (Y_2^2 - Y_4^2) = 0, \tag{9}$$

It can be put down in the form as

$$J(x) = x^2 + J_1 x + J_2 = 0 , \tag{*}$$

where $x = \xi^2$, and $J_1 = Y_1^2 - Y_3^2 - 2Y_2, J_2 = Y_2^2 - Y_4^2$

From above equation if $(A_3): Y_1^2 - Y_3^2 - 2Y_2 > 0, Y_2^2 - Y_4^2 > 0$

Therefore, if $Y_1^2 - Y_3^2 - 2Y_2 > 0 \text{ and } Y_2^2 - Y_4^2 > 0$ the equation (*) does not have positive roots.

The result established by the verification of conditions $(A_1),(A_2) \text{ and } (A_3)$ that all values of (6) have $\operatorname{Re}(\lambda) < 0$. It concluded by Rouche's results that the roots ξ of equation (9) have $\operatorname{Re}(\xi) < 0$. Thus, the positive steady point $E_2(q_1^*, q_2^*)$ for time delay $\tau > 0$ is locally asymptotically stable.

4.2.1 Hopf bifurcation

In the present subsection, we are proceeding to obtain conjugate imaginary roots which occur pairwise of equation (5).

If $(A_4): Y_2^2 - Y_4^2 < 0$, then by the results mentioned in principle of Routh-Hurwitz, the 4th order equation (9) will have positive root ξ_0 which exist uniquely. This further permit the Hopf bifurcation exist in the system (1) as a result of the condition that equation (5) will have a pair of complex roots $\pm i\xi_0$.

Put $\xi = \xi_0$ in real and imaginary part represented by equations (7) and (8) and simplify to obtain the value of τ as:

$$\tau_n^* = \frac{1}{\xi_0} \sin^{-1} \left\{ \frac{\xi_0 \left(\xi_0^2 Y_3 - Y_2 Y_3 + Y_1 Y_4 \right)}{Y_3^2 \xi_0^2 + Y_4^2} \right\} + \frac{2m\pi}{\xi_0} ; m = 0,1,2,3,.... \tag{10}$$

By using the result mentioned in Butler Lemma [20], the positive equilibrium point $E_2(q_1^*, q_2^*)$ rest stable for $\tau < \tau_0^*$ and for $m = 0$.

139

Also as discussed earlier, if (A_1) and (A_2) hold, then for $\tau = 0$, the equation (5) have negative real part of root. Hence, the positive equilibrium point $E_2(q_1^*, q_2^*)$ of the system (1) is conditionally stable.

In continuation, we focus on the process of Hopf bifurcation at positive steady point $E_2(q_1^*, q_2^*)$ when the bifurcation parameter which is taken to be time delay τ grows through the threshold value τ_0^*. By the theory of Hopf bifurcation [21, 22], there are two conditions for the occurrence of Hopf- bifurcation:

(i) A pair of complex conjugate purely imaginary latent values $\pm i\xi_0$ exists with the negative real part of all other latent values.

(ii) The transversality condition is satisfied.

The first condition for occurrence of Hopf bifurcation is already stated and proved in section 6. Next, we are going to check the condition of transversality when the real part of characteristic equation

(5) changes with τ close to τ_0^* i.e. $\left\{\dfrac{d}{d\tau}(\operatorname{Re}\lambda)\right\}_{\tau=\tau_0^*, \xi=\xi_0} > 0$

The periodic solution arises when the conditions of Hopf bifurcation are satisfied [23, 24]. Let $\lambda(\tau) = \mu(\tau) + i\xi(\tau)$ be the complex root of characteristic equation (5) near $\tau = \tau_0^*$ with the conditions that $\mu(\tau_0^*) = 0$ and $\xi(\tau_0^*) = \xi_0$. Substituting $\lambda(\tau)$ into (5), differentiate with τ, we get

$$sign\left\{\frac{d}{d\tau}(\operatorname{Re}\lambda)\right\}_{\tau=\tau_0^*, \xi=\xi_0} = sign\left\{\frac{\sqrt{(Y_3^2 - Y_1^2 + 2Y_2)^2 - 4(Y_2^2 - Y_4^2)}}{((-\xi_0^2 + Y_2)^2 + Y_1^2\xi_0^2)(Y_4^2 + Y_3^2\xi_0^2)}\right\}$$

(11)

By virtue of (A_4), $\left\{\dfrac{d}{d\tau}(\operatorname{Re}\lambda)\right\}_{\tau=\tau_0^*, \xi=\xi_0} > 0$. Thus the transversality

condition holds.

Based on above discussion, we can state another theorem as:

Theorem 7: [25] Assume that positive equilibrium point $E_2(q_1^*, q_2^*)$ exists and the conditions $Y_1^2 - Y_3^2 - 2Y_2 > 0, Y_2^2 - Y_4^2 < 0$ fulfilled for model defined by system of differential equations (1).

The necessary and sufficient conditions that the positive steady point $E_2(q_1^*, q_2^*)$ is L.A.S. with the association of time delay are

(i) The positive equilibrium point $E_2(q_1^*, q_2^*)$ is LAS in the interval $0 \le \tau < \tau_0^*$.

(ii) The positive equilibrium point $E_2(q_1^*, q_2^*)$ becomes unstable, when $\tau > \tau_0^*$

(iii) The Hopf bifurcation exists in the system (1) around positive equilibrium point $E_2(q_1^*, q_2^*)$ at critical value $\tau = \tau_0^*$. It gives $E_2(q_1^*, q_2^*)$ bifurcates into small periodic solutions when the value of τ crosses τ_0^*. The value of τ_0^* form $=0$

is given by $\tau_0^* = \dfrac{1}{\xi_0} \cdot \sin^{-1}\left\{ \dfrac{\xi_0 \left(\xi_0^2 Y_3 - Y_2 Y_3 + Y_1 Y_4 \right)}{Y_3^2 \xi_0^2 + Y_4^2} \right\}.$

5. Numerical simulations:

In this upcoming section, we perform numerical simulations to support our theoretical results. All the figures in this paper are drawn using MATLAB. The assumed set of parametric values for the system (1) is described as:

$a = 0.4, K = 2, \alpha = 3.7, \alpha_1 = 2.4, \beta = 12.2, \delta_1 = 0.2, \delta_2 = 0.15, \gamma = 3.2, \theta = 0.5$

Using these parametric values, the system of equations (1) becomes:

$$\frac{dq_1(t)}{dt} = (0.4)q_1\left(1 - \frac{q_1}{2}\right) - \frac{(3.7)q_1(t)q_2(t-\tau)}{12.2 + q_1(t)} - (0.2)q_1(t),$$

$$\frac{dq_2(t)}{dt} = \frac{(2.4)(3.7)q_1(t)q_2(t-\tau)}{12.2 + q_1(t)} - (0.15)q_2(t) - \frac{(0.5)q_1^2 q_2}{(3.2)^2 + q_1^2}. \quad (12)$$

The above set of parametric values with initial conditions $q_1(t) = 0.13$, $q_2(t) = 0.21$ and with the condition of no delay $(\tau = 0)$, the system converges to asymptotically stable positive steady point $E_2(0.2133, 0.5263)$, which is shown in Fig.1.

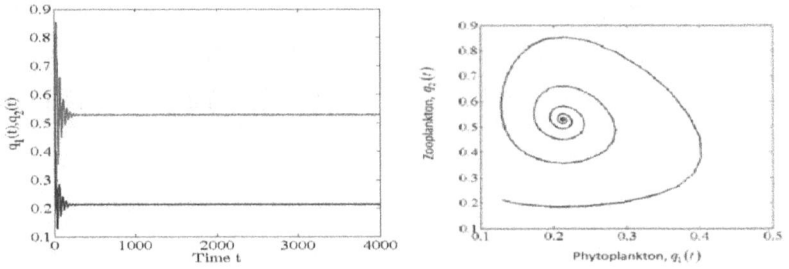

Fig.1. The trajectory of solution giving the local stability for $\tau = 0$ of phytoplankton and zooplankton population.

Using the software MATLAB, the system is solved numerically along with delay parameter τ and earns stability at $\tau = 0.95$ and this stable dynamical behaviour is depicted in Fig.2.

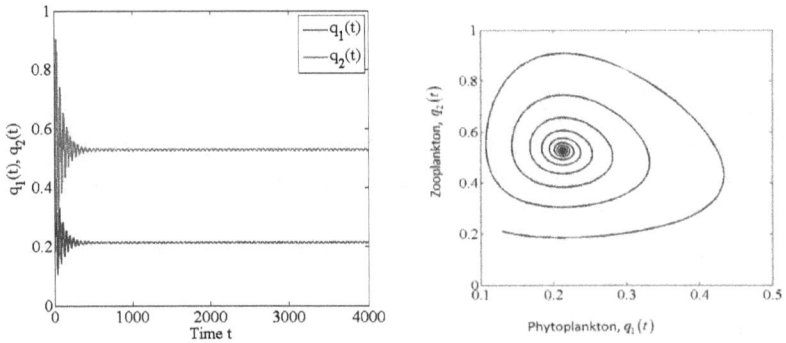

Fig.2. Convergence of solution curve to E_2 of the system (1) at $\tau = 0.95 < \tau_0^*$.

But when we fix all the parameters and give increment to the value of τ, we have found that small orbits occur in the system and Hopf bifurcation came to existence in the system. The phytoplankton and zooplankton population show oscillatory behaviour at $\tau = 3.2$ near point E_2, which is presented in Fig.3

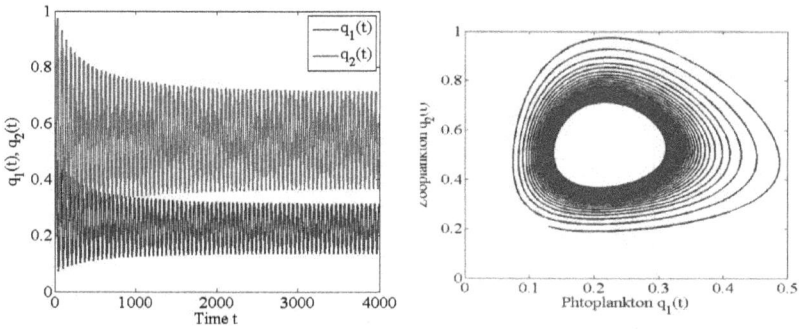

Fig.3. Phytoplankton and zooplankton population exhibit oscillatory motion around E_2 of the system (1) at $\tau = 3.2$.

The steady repeated solution of the system at $\tau = 4.2$ is shown in fig.4 and solution curve of the phytoplankton-zooplankton population around E_2 at $\tau = 12.2$ is shown in fig.5:

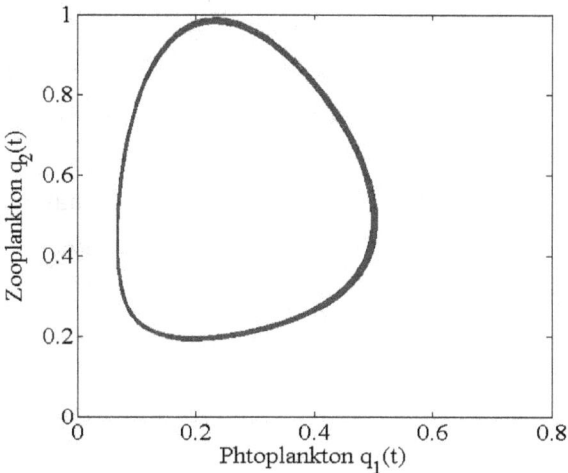

Fig.4. Stable periodic solution around E_2 for phytoplankton-zooplankton population at $\tau = 4.2$

143

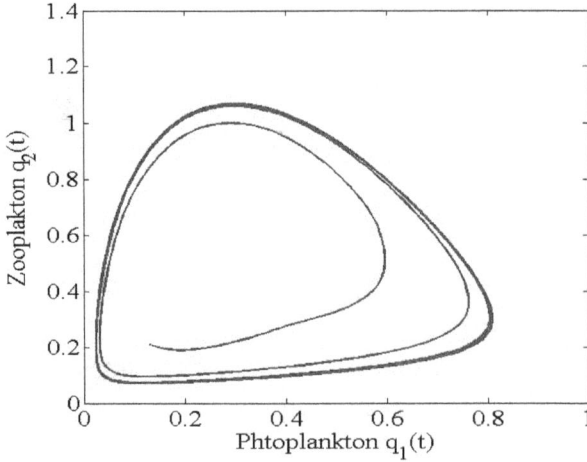

Fig.5. Solution curves around E_2 for phytoplankton-zooplankton population at $\tau = 12.2$

Numerically, the system is solved by considering the values of parameters as defined above, we find that the conditions $Y_1^2 - Y_3^2 - 2Y_2 = 0.00166 > 0$, $Y_2^2 - Y_4^2 = -0.000245 < 0$ are satisfied. By Routh-Hurwitz results, the equation (9) has positive root which exist uniquely, and thus a complex root $i\xi_0$, with $\xi_0 = 0.1219$ has been computed. Putting $\xi_0 = 0.1219$ in (9), we have $\tau_0^* = 3.085$, which is the threshold value of time delay for the system such that when the system crosses this critical value, it loses its stability. Moreover, the transversality condition $\left\{ \dfrac{d}{d\tau} (\text{Re}\,\lambda) \right\}_{\tau = \tau_0^*, \xi = \xi_0} > 0$ has been verified, which is the primary result for the occurrence of Hopf bifurcation.

6. Concluding remarks

Plankton are very important part of aquatic ecosystem. They are responsible for the generation of about 50% of oxygen in the atmosphere and help to absorb carbon emissions. They form the base of aquatic food webs as 95% of ocean life is made up of plankton. In the present chapter, the behaviour of phytoplankton and zooplankton population has been studied mathematically by taking into consideration the parameter of time delay τ as bifurcation parameter. The system is analysed for positive and bounded solution. The various steady state points along with their stability analysis have been discussed. The trivial steady state point $E_0(0,0)$ is stable if $a < \delta_1$. The conditions for which zooplankton free equilibrium point $E_1(q_1,0)$, where $q_1 = \dfrac{K(a-\delta_1)}{a}$ without any time delay is LAS and the positive equilibrium point $E_2(q_1^*,q_2^*)$ with no time delay is asymptotically stable has been studied. The threshold value of delay parameter τ defined as τ_0^* is calculated and showed that when $\tau > \tau_0^*$, the positive equilibrium point $E_2(q_1^*,q_2^*)$ becomes unstable. As a result of this, Hopf bifurcation occurs in the system. The transversality condition for occurrence of Hopf bifurcation is verified.

The trajectory of solution giving the local stability for $\tau = 0$ of phytoplankton and zooplankton population is shown in Fig.1. Convergence of solution curve to E_2 of the system (1) at $\tau = 0.95 < \tau_0^*$ is depicted in Fig.2. The phytoplankton and zooplankton populations exhibited oscillatory behaviour around E_2 of the system (1) at $\tau = 3.2$, which is presented in Fig.3. Stable periodic solution around E_2 for phytoplankton-zooplankton population at $\tau = 4.2$, is shown in Fig.4.

7. Acknowledgments

This research work is completed in the Numerical Analysis Laboratory of Shaheed Bhagat Singh State University, Ferozepur, Punjab (India). The authors admiringly recognize the assistance provided by the faculty and staff of University for providing the space to finish this study.

References

1. Upadhyay, R. K., Naji, R. K., Kumari, N.(2007) Dynamical Complexity in Some Ecological Models: Effect of Toxin Pro-

duction by Phytoplankton. *Nonlinear Analysis: Modelling and Control,* 12(1), 123–138.

2. Chattopadhayay, J., Sarkar, R.R., and Mandal, S. (2002) Toxin-producing plankton may act as a biological control for planktonic blooms field study and mathematical modelling. *Journal of Theoretical Biology,* 215(3), 333–344.

3. Chattopadhyay, J., Sarkar, R.R., and El Abdllaoui, A. (2002) A delay differential equation model on harmful algal blooms in the presence of toxic substances. *Mathematical Medicine and Biology: A Journal of the IMA,* 19(2), 137–161.

4. Saha, T., and Bandyopadhyay, M. (2009) Dynamical analysis of toxin producing phytoplankton–zooplankton interactions. *Nonlinear Analysis: Real World Applications,* 10(1), 314–332.

5. Zhao, Q., Liu, S., and Niu, X. (2020) Effect of water temperature on the dynamic behavior of phytoplankton-zooplankton model. *Applied Mathematics and Computation, Elsevier,* 378, 125211.

6. Rehim, M., and Imran, M. (2012) Dynamical analysis of a delay model of phytoplankton–zooplankton interaction. *Applied Mathematical Modelling,* 36(2), 638–647.

7. Jiang, Z., Ma, W., and Li, D. (2014) Dynamical behavior of a delay differential equation system on toxin producing phytoplankton and zooplankton interaction. *Japan Journal of Industrial and Applied Mathematics, Springer,* 31, 583-609.

8. Sharma, A., Sharma, A. K., and Agnihotri, K. (2015) Analysis of a toxin producing phytoplankton-zooplankton interaction with Holling IV type scheme and time delay. *Nonlinear Dynamics, Springer,* 81, 13-25.

9. Boonrangsiman, S., Bunwong, K., and Moore, E.J. (2016) A bifurcation path to chaos in a time-delay fisheries predator–prey model with prey consumption by immature and mature predators. *Mathematics and Computers in Simulation,* 124, 16–29.

10. Dai, C., Zhao, M., Yu, H., & Wang, Y. (2015) Delay-induced instability in a nutrient-phytoplankton system with flow. *Physical Review E, 91*(3), 032929.

11. Kumar, R. (2021) Mathematical modeling of a delayed innovation diffusion model with media coverage in adoption of an innovation. In *Systems Reliability Engineering,* De Gruyter, pp. 153-172.

12. Tiancai, L., Hengguo, Yu., Chuanjun, D., Min, Z. (2020) Impact of noise in a phytoplankton-zooplankton system. *Journal of Applied Analysis & Computation*, 10(5), 1878-1896.
13. Kumar. V., Kumari N., (2020) Stability and bifurcation analysis of Hassell-varley prey predator system with fear effect. *International journal of Applied and Computational Mathematics*, 150.
14. Kumar, R., Sharma, A. K., and Agnihotri, K. (2019) Bifurcation analysis of a nonlinear diffusion model: Effect of evaluation period for the diffusion of technology. *Arab Journal of Mathematical Sciences*, 25(2), 189-213.
15. Birkhoff, G., and Rota, G. (1989) *Ordinary differential equations*, Ginn, Boston.
16. Luenberger, D.G. (1979) *Introduction to dynamic systems; theory, models, and applications.*
17. Gopalsamy, K. (2013) *Stability and oscillations in delay differential equations of population dynamics*, vol. 74, Springer Science & Business Media.
18. Li, F., Li, H. (2012) Hopf bifurcation of a predator–prey model with time delay and stage structure for the prey. *Mathematical and Computer Modelling*, 55(3), 672–679.
19. Song, Y., Wei, J., Han, M. (2004) Local and global hopf bifurcation in a delayed hematopoiesis model. *International Journal of Bifurcation and Chaos*, 14(11), 3909–3919.
20. Freedman, H., and Waltman, P. (1985) Persistence in a model of three competitive populations. *Mathematical Biosciences* 73(1), 89–101.
21. Edelstein-Keshet, L. (1988) *Mathematical models in biology*, vol. 46. Siam.
22. Kuznetsov, Y.A. (2004) *Elements of Applied Bifurcation Theory*, Third Edition (Applied Mathematical Sciences), vol 112, New York: Springer Verlag.
23. Hale, J.K. (1969) Ordinary *Differential Equations*, Wiley, New York.
24. Kuang, Y. (1993) *Delay differential equations: with applications in population dynamics*, Academic Press.
25. Kar, T., Pahari, U. (2007) modelling and analysis of a prey-predator system with stage-structure and harvesting. *Nonlinear Analysis: Real World Applications*, 8(2), 601–609.

CHAPTER 9

Bifurcation Behaviors of a Phytoplankton-Zooplankton Model with Toxicity

Rakesh Kumar[1], Navneet Rana[2] and Vijay Kumar[3]
[1]Shaheed Bhagat Singh State University, Ferozepur, Punjab, India
[2]Guru Nanak College, Sri Muktsar Sahib, Punjab, India
[3]Sardar Beant Singh State University, Gurdaspur, Punjab, India

1. Introduction

Marine ecosystem commonly involves the contribution of Plankton. The marine food web is entirely based on plankton. The plankton mainly occurs in two categories, one is phytoplankton and other is zooplankton. The phytoplankton are plant form of plankton which are green in colour and zooplankton are the animal form of plankton. These are also termed as 'marine drifters' as they are free to move along the direction of current. There is very strong relationship exists between phytoplankton and zooplankton population in marine ecosystem. Phytoplankton prepares their food with the process of photosynthesis. The growth of phytoplankton population depends upon the nutrients available in water. Zooplankton eats phytoplankton as their food.

Plankton are responsible to produce about 50% of total oxygen in the atmosphere and thus help to maintain ecological balance. Phytoplankton also contributes to global warming as they absorb about 50% of the atmospheric carbon dioxide. When the concentration of nutrients available in water becomes very high, the phytoplankton species may grow in a large amount which results in harmful algal blooms (HABs). Some blooms occur due to change in water temperature and its saltiness. Such blooms produce poisonous chemicals in water. These toxin substances have negative impact on other aquatic organisms like zooplankton and then transferred to fishes and sooner or later affect the human beings. There are many factors which can be considered to learn about their interactions such as nutrients in water, temperature, toxicity, harvesting etc. Mathematical models have made remarkable contribution to recognize the driving force which relates the phytoplankton population and zoo-plankton population with toxicity [1-14]. J. Chattopadhyay [15] presented a mathematical model to show the interlinkage between toxic producing phytoplankton (TPP) and zooplankton by defining the model as

$$
\left.\begin{aligned}
\frac{dn_1}{dt} &= g\left[1 - \frac{n_1(t)}{K}\right]n_1(t) - \alpha\chi(n_1(t))n_2(t), \\
\frac{dn_2}{dt} &= \alpha_1\chi(n_1(t))n_2(t) - \delta n_2(t) - \theta\rho(n_1(t))n_2(t),
\end{aligned}\right\}
\tag{1}
$$

Here, $n_1(t)$ represents the phytoplankton population density which produces toxins and $n_2(t)$ represents the zooplankton population density at time t. In system (1), g represents rate at which the phytoplankton population density grows naturally. Authors in [12, 16, 17, 20] modified system (1) by establishing delay parameter as time delay or by considering values of $\chi(n_1)$ and $\rho(n_1)$. Hopf-bifurcation analysis is done by these researchers to observe that the stability of the system is related to the impact of different parameters involved in the system. The present chapter aims to learn the interaction between phytoplankton and zooplankton populations with toxicity and Holling type III functional response. The phytoplankton produces toxic substances in water. The toxin substances produced by phytoplankton are very harmful for other organisms and also for human beings. We analyze the dynamics of phytoplankton and zooplankton system for positive and bounded solution. The condition under which different equilibrium points exist and their stability analysis has been established. By demonstrating the characteristic equation of jacobian around positive equilibrium point, the condition for occurrence of Hopf bifurcation has been found. Numerical Simulations are done to verify the analytic results using MATLAB.

The present paper is arranged in different sections as follows. The mathematical model is formulated in second section. In the subsequent section, the positivity and boundedness of solution has been discussed. The equilibrium points and their stability analysis are presented in fourth section. Further in the next section, the conditions for existence of Hopf bifurcation have been derived. Numerical simulation has been completed to verify the conceptual results in the next section. At last, the concluding remark have been given to elaborate the results.

2. Mathematical settings of the system

Let $n_1(t)$ and $n_2(t)$ be the phytoplankton population density and zooplankton population density respectively at any time t. The values of functions are taken as: $\chi(n_1) = \dfrac{n_1(t)}{\beta + n_1(t)}$ and $\rho(n_1) = \dfrac{n_1^2(t)}{\gamma^2 + n_1^2(t)}$. Then, the basic mathematical model to represent the given system of phytoplankton and zooplankton can be represented by following set of differential equations as:

$$\left.\begin{array}{l} \dfrac{dn_1(t)}{dt} = g\left[1 - \dfrac{n_1(t)}{K}\right]n_1(t) - \dfrac{\alpha n_1(t)n_2(t)}{\beta + n_1(t)} - hn_1(t), \\[3mm] \dfrac{dn_2(t)}{dt} = \dfrac{\alpha_1 n_1(t)n_2(t)}{\beta + n_1(t)} - \delta Z(t) - \dfrac{\theta n_1^2(t)n_2(t)}{\gamma^2 + n_1^2(t)}, \end{array}\right\} \tag{2}$$

The system as described in (2) is studied with conditions defined initially as $n_1(0)>0$, $n_2(0)>0$.

Table 1: Explanation of Parameters

Parameter	Description of Parameter
g	Phytoplankton intrinsic growth rate
K	Phytoplankton carrying capacity
α	Maximum uptake rate
α_1	Phytoplankton conversion rate to zooplankton
β	Half saturation constant
h	Natural mortality rate of phytoplankton
γ	Half saturation constant
θ	Rate at which toxic are released by phytoplankton
δ	Rate at which zooplankton die naturally

3. Analysis for positive solution and its Boundedness

In present section, we are performing calculations to describe that every finding as solution of the system is positive and bounded. As we are dealing with a model of biological existence, both the population densities exist positively and there are some limits or boundary points to the solutions. Thus, we can say that this system of equations possess positive and bounded solutions.

Theorem 1: Every solution of the mathematical model given in equation (2) is positive.
Proof: Rearrange the 1ˢᵗequation of system (2), and write it as

$$\frac{dn_1}{n_1} = \left\{ g\left(1 - \frac{n_1}{K}\right) - \frac{\alpha n_2}{\beta + n_1} - h \right\} dt,$$

$$\Rightarrow \frac{dn_1}{n_1} = \eta(n_1, n_2) dt,$$

where $\eta(n_1, n_2) = g\left(1 - \frac{n_1}{K}\right) - \frac{\alpha n_2}{\beta + n_1} - h.$

Integrate the equation from 0 to t, we get

$$\Rightarrow n_1(t) = n_1(0) \exp\left\{ \int_0^t \eta(n_1, n_2) dt \right\} > 0 \ \forall \ t.$$

Consider 2ⁿᵈequation defined in the system (2) of equations, we have

$$\frac{dn_2}{n_2} = \left\{ \frac{\alpha_1 n_1}{\beta + n_1} - \delta - \frac{\theta n_1^2}{\gamma^2 + n_1^2} \right\} dt,$$

$$\Rightarrow \frac{dn_2}{n_2} = \xi(n_1, n_2) dt,$$

where $\xi(n_1, n_2) = \frac{\alpha_1 n_1}{\beta + n_1} - \delta - \frac{\theta n_1^2}{\gamma^2 + n_1^2}.$

Integrate the equation within limits 0 to t, we get

$$n_2(t) = n_2(0) \exp\left\{ \int_0^t \xi(n_1, n_2) dt \right\} > 0 \ \forall \ t.$$

Thus we get that both the population densities are positive i.e. $n_1(t) > 0$ and $n_2(t) > 0$ $\forall$t.

The boundedness of result is an important part to discuss the behavior of system as it signifies that no population grows indefinitely.

Lemma 1:- (Comparison Lemma)Let us suppose that u, v > 0, with the condition f (0) is positive.

Then, for $\dfrac{df}{dt} \le \left(u - vf\left(t\right)\right)$, $\limsup\limits_{t \to \infty} f\left(t\right) \le \dfrac{u}{v}$ and also for

$\dfrac{df}{dt} \ge \left(u - vf\left(t\right)\right)$, $\liminf\limits_{t \to \infty} f\left(t\right) \ge \dfrac{u}{v}$.

Theorem-2:- The solutions of the mathematical model defined in equation (2) are always uniformly bounded.
Proof: The 1ˢᵗequation of system of equations (2) can be rearranged to get,

$$\dfrac{dn_1}{dt} \le g\left(1 - \dfrac{n_1(t)}{K}\right)n_1(t) - hn_1(t),$$

$$\Rightarrow \dfrac{dn_1}{dt} \le gn_1 - \dfrac{gn_1^2}{K} - hn_1,$$

UsingComparison Lemma 1, we can get the result as

$$Lim\,Sup\limits_{t \to \infty} n_1(t) \le \dfrac{(g-h)K}{g} = P_1\left(say\right).$$

Next, Let us focus on time dependent function $\Omega(t) = \alpha_1 n_1 + \alpha n_2$.
We can have

$$\dfrac{d\Omega}{dt} = \alpha_1 \dfrac{dn_1}{dt} + \alpha \dfrac{dn_2}{dt},$$

$$\Rightarrow \dfrac{d\Omega}{dt} = \alpha_1 gn_1 - \alpha_1 \dfrac{n_1^2 g}{K} - \alpha_1 hn_1 - \delta\alpha n_2 - \dfrac{\alpha\theta n_1^2 n_2}{\gamma^2 + n_1^2},$$

$$\Rightarrow \dfrac{d\Omega}{dt} = \alpha_1 gn_1 - \alpha_1 hn_1 - \delta\alpha n_2 - n_1^2\left(\alpha_1 \dfrac{g}{K} + \dfrac{\alpha\theta n_2}{\gamma^2 + n_1^2}\right),$$

$$\Rightarrow \dfrac{d\Omega}{dt} \le \alpha_1 n_1\left(g - h\right) - \alpha\delta n_2,$$

$$= \left(g - h + \delta\right)\alpha_1 n_1 - \delta\alpha_1 n_1 - \alpha\delta n_2,$$

$$\Rightarrow \frac{d\Omega}{dt} \leq W - \delta\left(\alpha_1 n_1 + \alpha n_2\right),$$

$$\Rightarrow \frac{d\Omega}{dt} + \Omega\delta \leq W, \text{ where } W = \left(g - h + \delta\right)\alpha_1 \frac{\left(g-h\right)K}{g}.$$

By calculations [18], we get $Lim\,Sup\ \underset{t\to\infty}{\Omega(t)} \leq \frac{W}{\delta} = P_2\,(say)$.

Therefore, model (2) has all the solutions as ultimately bounded.

4. Stability conditions of distinct equilibrium points

In the upcoming section, we are exploring the condition for the existence of various equilibrium points along with their steadiness. The equilibrium points for model (2) can be considered as follows:

(i) Trivial equilibrium point $E_0(0,0)$, which is due to the total disappearance of phytoplankton, zooplankton.

(ii) The equilibrium point $E_1\left(\dfrac{K(g-h)}{g},0\right)$, which is due to the disappearance of zooplankton. The zooplankton free equilibrium point $E_1\left(\dfrac{K(g-h)}{g},0\right)$ exist if g> h.

(iii) The positive state equilibrium point $E_2(n_1^*,n_2^*)$, where n_1^* is zero of cubic equation $\Delta_3 n_1^3 + \Delta_2 n_1^2 + \Delta_1 n_1 + \Delta_0 = 0$, (3)

where $\Delta_3 = \alpha_1 - \delta - \theta,\ \Delta_2 = -(\delta+\theta)\beta,\ \Delta_1 = (\alpha_1 - \delta)\gamma^2,$
$\Delta_0 = -\delta\beta\gamma^2.$

and $n_2^* = \dfrac{(\beta+n_1^*)(K(g-h)-gn_1^*)}{\alpha K}$

As $\Delta_0 < 0, \Delta_1 > 0, \Delta_2 < 0, \Delta_3 > 0$ the equation (3) has unique positive root if $\Delta < 0$ and the equation (3) has three distinct positive root if $\Delta > 0$, where $\Delta = 18\Delta_3\Delta_2\Delta_1\Delta_0 - 4\Delta_2^3\Delta_0 + \Delta_2^2\Delta_1^2 - 4\Delta_3\Delta_1^3 - 27\Delta_3^2\Delta_0^2.$
We use the Jacobian matrix in order to study the behavior of solution near all the equilibrium points. The Jacobian matrix calculated at any point (n_1, n_2) is given by

$$J_E = \begin{pmatrix} a_{11} & a_{12} \\ a_{21} & a_{22} \end{pmatrix}$$

where

$$a_{11} = g - \frac{2gn_1}{K} - \frac{\alpha\beta n_2}{(\beta + n_1)^2} - h;$$

$$a_{12} = -\frac{\alpha n_1}{\beta + n_1};$$

$$a_{21} = \frac{\alpha_1 \beta n_2}{(\beta + n_1)^2} - \frac{2\theta\gamma^2 n_1 n_2}{(\gamma^2 + n_1^2)^2};$$

$$a_{22} = \frac{\alpha_1 n_1}{\beta + n_1} - \delta - \frac{\theta n_1^2}{\gamma^2 + n_1^2}.$$

4.1 Stability Analysis

Next, in this part, the equilibrium points are analyzed for their stability. The model system is linearized throughout the corresponding equilibrium point to get the criterion for local stability of the equilibrium points.

Theorem 3:

(i) The trivial equilibrium point $E_0(0,0)$ is saddle point.

(ii) The zooplankton free equilibrium point $E_1\left(\dfrac{K(g-h)}{g}, 0\right)$ is

        locally   asymptotically   stable   if

$$\frac{\alpha_1 K(g-h)}{\beta g + K(g-h)} < \delta + \frac{\theta K^2 (g-h)^2}{\gamma^2 g^2 + K^2 (g-h)^2}.$$

(iii) The positive state equilibrium point $E_2(n_1^*, n_2^*)$ is locally asymptotically stable if $Tr < 0$ and $Det > 0$.

Proof:-

(i)   The Jacobian matrix calculated at trivial equilibrium point $E_0(0,0)$ can be defined as

$$J_{E_0} = \begin{pmatrix} g-h & 0 \\ 0 & -\delta \end{pmatrix}$$

The Jacobian matrix J_{E_0} has eigenvalues as $g-h, -\delta$. For $g > h$, the first eigenvalue is positive and second eigenvalue is negative. So, the equilibrium point $E_0(0,0)$ is saddle point.

(ii) The zooplankton free equilibrium point $E_1\left(\dfrac{K(g-h)}{g}, 0\right)$ has Jacobian matrix given by

$$J_{E_1} = \begin{pmatrix} b_{11} & b_{12} \\ b_{21} & b_{22} \end{pmatrix} \text{where}$$

$$b_{11} = -(g-h), b_{12} = -\frac{\alpha K(g-h)}{\beta g + K(g-h)}, b_{21} = 0, b_{22} = \frac{\alpha_1 K(g-h)}{\beta g + K(g-h)} - \delta - \frac{\theta K^2 (g-h)^2}{\gamma^2 g^2 + K^2 (g-h)^2}.$$

The Jacobian matrix J_{E_1} has eigenvalues as $-(g-h)$ and

$$\frac{\alpha_1 K(g-h)}{\beta g + K(g-h)} - \delta - \frac{\theta K^2 (g-h)^2}{\gamma^2 g^2 + K^2 (g-h)^2}. \text{ For } g > h, \text{ the first eigen}$$

value is negative.

Hence, $E_1\left(\dfrac{K(g-h)}{g}, 0\right)$ is locally asymptotically stable if

$$\frac{\alpha_1 K(g-h)}{\beta g + K(g-h)} - \delta - \frac{\theta K^2 (g-h)^2}{\gamma^2 g^2 + K^2 (g-h)^2} < 0$$

i.e. $\dfrac{\alpha_1 K(g-h)}{\beta g + K(g-h)} < \delta + \dfrac{\theta K^2 (g-h)^2}{\gamma^2 g^2 + K^2 (g-h)^2}.$

(iii) The Jacobian matrix at positive state equilibrium point $E_2(n_1^*, n_2^*)$ is $J_{E_2} = \begin{pmatrix} c_{11} & c_{12} \\ c_{21} & c_{22} \end{pmatrix}$ where

$$c_{11} = g - \frac{2gn_1^*}{K} - \frac{\alpha\beta n_2^*}{\left(\beta + n_1^*\right)^2} - h;$$

$$c_{12} = -\frac{\alpha n_1^*}{\beta + n_1^*};$$

$$c_{21} = \frac{\alpha_1\beta n_2^*}{\left(\beta + n_1^*\right)^2} - \frac{2\theta\gamma^2 n_1^* n_2^*}{(\gamma^2 + n_1^{*2})^2};$$

$$c_{22} = \frac{\alpha_1 n_1^*}{\beta + n_1^*} - \delta - \frac{\theta n_1^{*2}}{\gamma^2 + n_1^{*2}}.$$

The Jacobian matrix J_{E_2} has characteristic equations $(\lambda^2 - Tr\lambda + Det) = 0$, where

$$Tr = c_{11} + c_{22}$$

$$= g - \frac{2gn_1^*}{K} - \frac{\alpha\beta n_2^*}{\left(\beta + n_1^*\right)^2} - h + \frac{\alpha_1 n_1^*}{\beta + n_1^*} - \delta - \frac{\theta n_1^{*2}}{\gamma^2 + n_1^{*2}}; \text{ and}$$

$$Det = c_{11}c_{22} - c_{12}c_{21}$$

$$= \left[g - \frac{2gn_1^*}{K} - \frac{\alpha\beta n_2^*}{\left(\beta + n_1^*\right)^2} - h \right]\left[\frac{\alpha_1 n_1^*}{\beta + n_1^*} - \delta - \frac{\theta n_1^{*2}}{\gamma^2 + n_1^{*2}} \right] - \left[\frac{\alpha n_1^*}{\beta + n_1^*} \right]\left[\frac{\alpha_1\beta n_2^*}{\left(\beta + n_1^*\right)^2} - \frac{2\theta\gamma^2 n_1^* n_2^*}{(\gamma^2 + n_1^{*2})^2} \right];$$

Using the Routh-Hurwitz condition [19], the characteristic equation will have either real roots which are negative or complex conjugate roots which occur pairwise and have negative real part if *Trace of matrix* < 0 *and* Determinant *of matrix* > 0. The criteria for existence of different equilibrium points along with their stability criteria can be summarized in Table 2 as:

Table 2: Criteria for existence of various equilibrium points and their stability conditions

Sr	Equilibrium Points	Criteria for Existence	Criteria for Stability
1.	$E_0(0,0)$	Always Exist	Saddle Point
2.	$E_1\left(\dfrac{K(g-h)}{g},0\right)$	$g > h$	$\dfrac{\alpha_1 K(g-h)}{\beta g + K(g-h)} < \delta + \dfrac{\theta K^2(g-}{\gamma^2 g^2 + K^2(}$

3.	$E_2(n_1^*, n_2^*)$	As $\Delta_0 < 0, \Delta_1 > 0, \Delta_2 <$ unique positive root if $\Delta < 0$ three distinct positive root if $\Delta > 0$,	$Tr < 0$ and Det >0

5. Hopf bifurcation

In this present section, we are going to examine the condition of the Hopf bifurcation. Hopf bifurcation is that particular value at which the system changes its stability and this result into periodic solution. This particular value at which the stability switches is called Hopf bifurcation point. The Jacobian matrix at positive state equilibrium point $E_2(n_1^*, n_2^*)$ is

$$J_{E_2} = \begin{pmatrix} c_{11}(g) & c_{12}(g) \\ c_{21}(g) & c_{22}(g) \end{pmatrix} \tag{4}$$

where

$$c_{11} = g - \frac{2gn_1^*(g)}{K} - \frac{\alpha\beta n_2^*(g)}{\left(\beta + n_1^*(g)\right)^2} - h; \; c_{12} = -\frac{\alpha n_1^*(g)}{\beta + n_1^*(g)}; \; c_{21} = \frac{\alpha_1\beta n_2^*(g)}{\left(\beta + n_1^*(g)\right)^2} - \frac{2\theta\gamma^2 n_1^*(g)n_2^*(g)}{(\gamma^2 + n_1^{*2}(g))^2};$$

$$c_{22} = \frac{\alpha_1 n_1^*(g)}{\beta + n_1^*(g)} - \delta - \frac{\theta n_1^{*2}(g)}{\gamma^2 + n_1^{*2}(g)}.$$

The Jacobian matrix J_{E_2} has characteristic equation as

$$(\lambda^2 - Tr(g)\lambda + Det(g)) = 0 \tag{5}$$

where

$Tr(g) = sum \; of \; diagonal \; entries = c_{11}(g) + c_{22}(g),$

$Det(g) = c_{11}(g)c_{22}(g) - c_{12}(g)c_{21}(g).$

The characteristic equation (5) must have eigenvalues pairwise as

$$\lambda_{1,2} = \frac{Tr(g) \pm \sqrt{Tr^2(g) - 4Det(g)}}{2}$$

For Hopf bifurcation to exist, the Jacobian matrix (4) should have purely imaginary eigenvalues. Let $Tr(g) = 0$ for some g and $Det(g) > 0$ for any value of g. So, the characteristic equation (5) at critical value $g = g^*$ becomes

$$\lambda^2 + Det(g) = 0 \qquad\qquad (6)$$

The equation (6) must have $\lambda_{1,2} = \pm ib_0$, where $b_0 = \sqrt{Det(g)}$

Next, we show the condition of transversality [21, 22]. Let in the neighbourhood of $g = g^*$, at any point g we have

$$\lambda_{1,2} = \frac{p(g) \pm iq(g)}{2}, \text{ where } p(g) = \frac{Tr(g)}{2}, q(g) = \frac{\sqrt{4Det(g) - Tr^2(g)}}{2}.$$

Then,   the   transversality   condition

$$\left. \frac{d}{dg} p(g) \right|_{r=r^*} = \frac{1}{2} \left[c_{11}(g) + c_{22}(g) \right]_{g=g^*} = \frac{1}{2} Tr(g) \Big|_{g=g^*} \neq 0 \text{ holds and we}$$

can say that the system of equations (2) experiences Hopf bifurcation at the crucial value $g = g^*$.

6. Numerical Simulation

The following section deals with the numerical simulation of dynamics of phytoplankton and zooplankton model with toxicity. For this, we consider the value of parameters as:

$$K = 15, \alpha = 7.8, \alpha_1 = 4.1, \beta = 8.2, \gamma = 15.2, \delta = 0.21, \theta = 0.5, h = 0.32;$$

So, the system (2) for these parametric values takes the form as:

$$\left. \begin{aligned} \frac{dn_1}{dt} &= gn_1(t)\left[1 - \frac{n_1(t)}{(15)}\right] - \frac{(7.8)n_1(t)n_2(t)}{(8.2) + n_1(t)} - (0.32)n_1(t), \\ \frac{dZ}{dt} &= \frac{(4.1)n_1(t)n_2(t)}{(8.2) + n_1(t)} - (0.21)n_2(t) - \frac{(0.5)n_1^2(t)n_2(t)}{(15.2)^2 + n_1^2(t)}, \end{aligned} \right\} \qquad (7)$$

Consider the initial values of population density of phytoplankton and population density of zooplankton as $n_1(t) = 0.3$, and $n_2(t) = 0.61$ respectively. By giving consideration to these initial values of all the parameters, the system is integrated with respect to natural growth rate g with the help of MATLAB DDE23 pack. The LAS solution of system about positive equilibrium point $E_2(0.4436, 0.1831)$ at $g = 0.5$, which is clearly depicted in fig.1.

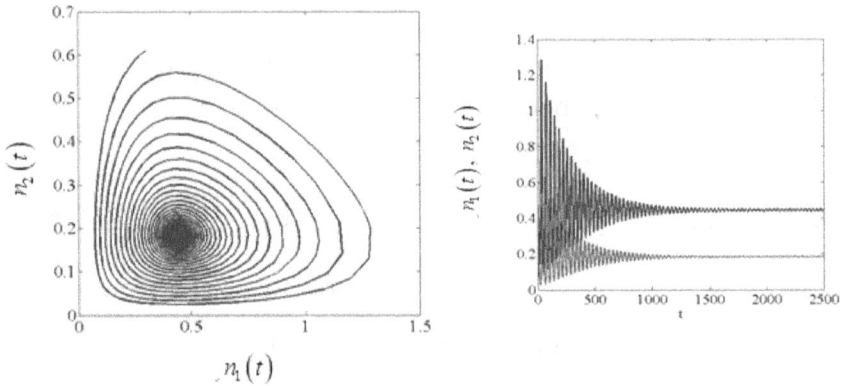

Fig. 1: The LAS solution of system about positive equilibrium point $E_2\left(0.4436, 0.1831\right)$ at $g = 0.5$.

When we moderately raise the value of g by considering all other specifications to be same as defined above, the system possesses oscillatory behavior and Hopf bifurcation aroused in the system at $g = 0.75$, which is demonstrated in fig.2:

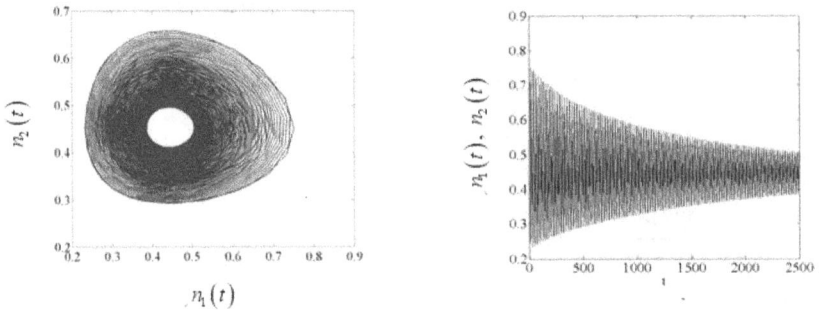

Fig. 2: Hopf bifurcation aroused in the system at $g = 0.75$.

When we further increase the value of $g = 3.25$, the natural rate at which the phytoplankton population density grows, then the system possesses periodic orbits around equilibrium point E_2 as shown in fig.3:

160

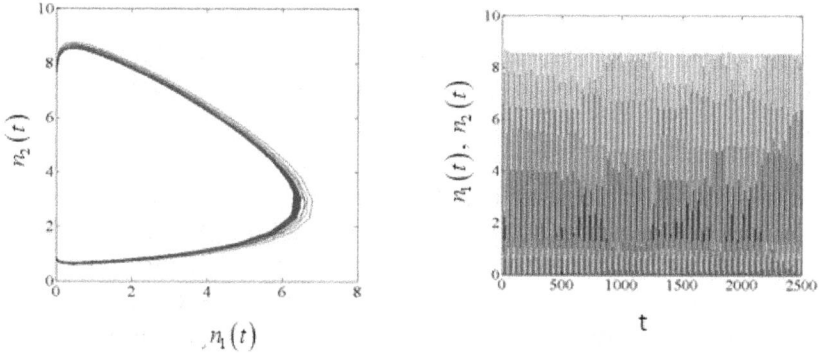

Fig. 3: Periodic orbital motion around positive state equilibrium point E_2

Numerically, the set of values for specifications as defined earlier in this section can be used to find $c_{11} = -0.00116, c_{12} = -0.4003, c_{21} = 0.2025, c_{22} = 0.000022,$
Also, $Tr = -0.001138 < 0$ and $Det = 0.0811 > 0$. These are the required conditions for applying the Routh-Hurwitz criterion. Thus, the roots of characteristic equation has two categories, either the real roots exist which are negative or complex conjugate root which occur pairwise with negative real part.
Also,

$$\frac{d}{dg}p(g)\bigg|_{g=0.75} = \frac{1}{2}[c_{11}(g) + c_{22}(g)]_{g=0.75} = \frac{1}{2}Tr(g)\bigg|_{g=0.75} = \frac{1}{2}(-0.001138) = -0.00057 \neq 0$$

This shows that the condition of transversality for existence of Hopf bifurcation is verified.

7. Concluding Remarks

Plankton is important part of aquatic ecosystem. The relationship between phytoplankton and zooplankton are of great importance. Many mathematical models are suggested to discuss the interaction between phytoplankton population and zooplankton population. Many members of phytoplankton population have a remarkable feature of formation of rapid and massive blooms. These members of phytoplankton community release poisonous chemicals which results in formation of algal blooms. These harmful algal blooms are natural toxins which are also very dangerous for other organisms

161

and also to the humans. Hence, it is very essential requirement to study marine ecology by using mathematical models which involves the active behavior of toxin producing phytoplankton and zooplankton.

In the present paper, such mathematical model is established to learn the active relation between phytoplankton and zooplankton with toxicity. The system has been investigated for positive solution and its boundedness. All the equilibrium points are analyzed for their stability conditions. The trivial equilibrium point is saddle point for $g > h$, the equilibrium point having no zooplankton population density i.e. $E_1 \left(\dfrac{K(g-h)}{g}, 0 \right)$ is stable for $b_{22} < 0$. The positive state equilibrium point $E_2 \left(n_1^*, n_2^* \right)$ is stable for $Tr < 0$ and $Det > 0$.

The transversality condition of Hopf bifurcation is verified, which ensures that the Hopf bifurcation occurred in the system w.r.t. bifurcation parameter g.

The LAS solution of system about positive equilibrium point $g = 0.5$, which is displayed in fig.1. When we moderately rise the value of g by fixing all other parameters, the Hopf bifurcation aroused in the system at $g = 0.75$, which is demonstrated in fig.2. The periodic orbits in the system are shown at $g = 3.25$ in fig. 3. All the results are verified by using numerical simulation.

Acknowledgements
This research work is completed in the Numerical Analysis Laboratory of Shaheed Bhagat Singh State University, Ferozepur, Punjab (India). The authors admiringly recognize the assistance provided by the faculty and staff of the University for providing the space to finish this study.

References

1. Anderson, D. (1989) Toxic algae blooms and red tides: A global perspective. *Red Tides: Biology, Environmental Science and Toxicology* (Elsevier), pp.11-21.

2. Hallegraeff, G. (1993) A review of harmful algae blooms and the apparent global increase. *Phycologia, 32*, 79– 99.
3. Roy, S., Alam, S. & Chattopadhyay, J. (2006) Competitive effects of toxin-producing phytoplankton on overall plankton populations in the Bay of Bengal. *Bulletin of Mathematical Biology, 68*, 2303–2320.
4. Roy, S., Bhattacharya, S., Das, P. & Chattopadhyay, J. (2007) Interaction among non-toxic phytoplankton, toxic phytoplankton and zooplankton inferences from field observations. *Journal of Biological Physics, 33*, 1–17.
5. Sarkar, R., Mukhopadhyay, B., Bhattacharyya, R. & Banerjee, S. (2007) Time lags can control algal bloom in two harmful phytoplankton–zooplankton system. *Applied Mathematicsand Computation, 186*, 445–459.
6. Saha, T. & Bandyopadhyay, M. (2009) Dynamical analysis of toxin producing phytoplankton–zooplankton interactions. *Nonlinear Analysis: Real World Applications, 10*, 314–332.
7. Wang, Y., Wang, H. & Jiang, W. (2014) Hopf transcritical bifurcation in toxic phytoplankton– zooplankton model with delay. *Journal of Mathematical Analysis and Applications, 415*, 574–594.
8. Sharma, A., Sharma, A. & Agnihotri, K. (2015) Analysis of a toxin producing phytoplankton–zooplankton interaction with Holling IV type scheme and time delay. *Nonlinear Dynamics, 81*, 1–13.
9. Sharma, A., Sharma, A. & Agnihotri, K. (2016) Spatio-temporal dynamic of toxin producing phytoplankton (TPP)-zooplankton interaction. *International Journal of Mathematical Modelling and Computations, 6*, 189–197.
10. Jiang, Z., Zhang, W., Zhang, J., Zhang, T., (2018) Dynamical Analysis of a phytoplankton-zooplankton system with harvesting term and Holling III functional response. *International Journal of bifurcation and Chaos, 28*, 1850162.
11. Jiang, Z., Bi, X., Zhang, T., Sampath Aruna Pradeep, B.G., (2019) Global Hopf bifurcation of a delayed phytoplankton-zooplankton system considering toxin producing effect and delay dependent coefficient. *Mathematical biosciences and engineering, 16*(5), 3807-3829.
12. Jiang, Z., Ma, W. & Li, D. (2014) Dynamical behavior of a delay differential equation system on toxin producing phyto-

plankton and zooplankton interaction. *Japan Journal of Industrial and Applied Mathematics*, 31, 583–609.

13. Jiang, Z. & Zhang, T. (2017) Dynamical analysis of a reaction-diffusion phytoplankton–zooplankton system with delay. *Chaos Solitons Fractals*, 104, 693– 704.

14. Khare, S., Misra, O. P., Dhar, J., (2010) Role of toxin producing phytoplankton on a plankton ecosystem. *Nonlinear Analysis: Hybrid Systems*, 496-502.

15. Chattopadhyay, J. & Sarkar, R. (2002) A delay differential equation model on harmful algal blooms in the presence of toxic substances. *IMA Journal of Mathematics Applied in Medicine and Biology*, 19, 137–161.

16. Rehim, M. & Imran, M., (2012) Dynamical analysis of a delay model of phytoplankton-zooplankton interaction. *Applied Mathematical Modelling*, 36, 638–647.

17. Wang, Y., Jiang, W., Wang, H. (2013) Stability and global Hopf bifurcation in toxic phytoplankton zooplankton model with delay and selective harvesting. *Nonlinear Dynamics*, 73, 881-896.

18. Birkhoff, G., and Rota, G. (1989) *Ordinary differential equations*, Ginn, Boston.

19. Luenberger, D.G.D.G. (1979) *Introduction to dynamic systems; theory, models, and applications.*

20. Kumar, V., & Kumari. N., (2020) Stability and bifurcation analysis of Hassell-varley prey predator system with fear effect. *International journal of Applied and Computational Mathematics*, 150

21. Hale, J., & Lunel, S., (1993) *Introduction to Functional Differential Equations*, Springer-Verlag, New York.

22. Hassard, B., Kazarinoff, N., & Wan, Y., (1981) Theory and Application of Hopf Bifurcation, *Cambridge University Press, Cambridge.*

CHAPTER 10

Stability analysis of crop-pest-natural enemy system as biological control approach

Vijay Kumar[1], Rakesh Kumar[2], Rishi Tuli[3], Alka[4]
[1]Sardar Beant Singh State University, Gurdaspur, Punjab, India
[2]Shaheed Bhagat Singh State University, Ferozepur, Punjab, India
[3]Sardar Beant Singh State University, Gurdaspur, Punjab, India
[4]Government Girls Sr. Sec. School, Dinanagar, Gurdaspur, Punjab, India

1. Introduction

As pests are harmful to plants and their control has become a major issue in agriculture. Outbreaks of insects appear to be exacerbated by some basic environmental problems [1, 2]. Usually, pesticides are used to lower the population of pests, which quickly eradicate these species, however, these pests are dangerous not only for humans now but also for agriculture [3]. Therefore, to assist people and agribusiness, biologically, that is, to save the climate naturally, steps have already been taken in this regard by authors, economists etc., so organic pest management strategies are safe. For instance, researchers have proposed models with a dietary supplement for predators to bring down the number of bugs [4, 5, 6, 7, 9, 10, 11, 14]. Thus, a biological control technique is employed to control the pest population using a pest-natural enemy compartment system.

Because of the above literature survey here, we propose a compartmental system, by considering the phenomenon that the natural systems control harmful pests biologically. The next sections of the model are, the proposed model, the concept of positivity and boundedness, the stability analysis, the sensitivity analysis, simulations and conclusions.

We assume that if the system is insect-free, a particular plant species grow naturally, and the insects have other food to survive, detailed modelling assumptions are as follows:

2. The proposed model

(i) Species are classified into three types, namely, the plant type U (t), the pest type V (t), and the natural enemy type W (t).

(ii) Let Λ be the plant's natural growth rate, without pests in the system.

(iii) Suppose the predation rate by pests of plants is h_1. Let the predation rate by natural enemies of pests be h_2.

(iv) For insects and natural enemies, let c_1 and c_2 represent the conversion rates respectively. Also, the death rates for the plants, the pests and the natural enemies, are d_1, d_2 and d_3 respectively.

$$\frac{dU}{dt} = \Lambda - h_1 UV - d_1 U,\tag{1}$$

$$\frac{dV}{dt} = c_1 UV - h_2 VW - d_2 V,\tag{2}$$

$$\frac{dw}{dt} = c_2 VW - d_3 W,\tag{3}$$

Assume that $U > 0$, $V > 0$ and $W > 0$ for t=0.

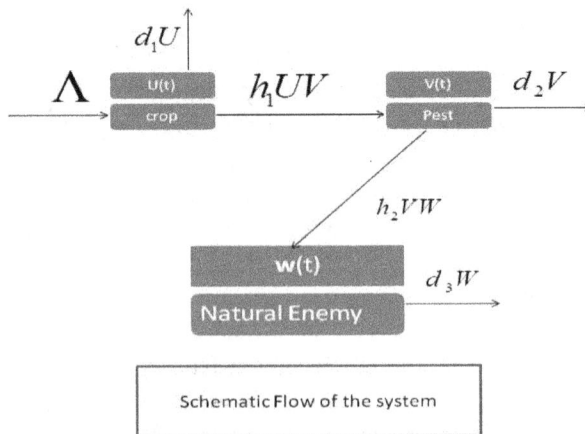

Schematic Flow of the system

166

3. The concept of positivity and boundedness

Presented below are some lemmas for positive and bounded solutions to the proposed systems (1)-(3):

3.1 Lemma When the initial populations are non-negative, the solutions of said system become non-negative $\forall\, t \geq 0$.

Proof: In the proposed system (1)-(3) with non-negative initial populations, let (U(t), V(t), and W(t)) represent the solutions. It is possible to write equation (1) as follows for all t,

$\dfrac{dU}{dt} \geq -h_1 UV - d_1 U$, it follows that $U(t) \geq U(0)\, e^{-\int_0^t (d_1 + h_1 V)dU} > 0$.

Also, for all t, the equation (2) can be represented as,

$\dfrac{dV}{dt} \geq -h_2 VW - d_2 V$, which evidences that

$V(t) \geq V(0)\, e^{-\int_0^t (d_2 + h_2 W)dU} > 0$.

Moreover, from the equation (3), we have for all t, $\dfrac{dw}{dt} \geq -d_3 W$,

which results that, $W(t) \geq W(0)\, e^{-\int_0^t (d_3)du} > 0$.

Thus, the proposed system is non-negative.

3.2 Lemma Presented below the uniformly bound in Ω, for the solutions to the proposed systems (1)-(3):

$$\Omega = \left\{ (U,V,W) : 0 \leq H(t) \leq \frac{\Lambda}{d'} \right\}, \quad d' = \min\{d_1, d_2, d_3\}.$$

Proof: Let us assume that $H = U + V + W$. Differentiating H w.r.t. t, we get,

$$\frac{dH}{dt} = \Lambda - h_1 UV - d_1 U + c_1 UV - h_2 VW - d_2 V + c_2 VW - d_3 W.$$

Since $c_1 \ll h_1$, $c_2 \ll h_2$ therefore we have,

$$\frac{dH(t)}{dt} = \Lambda - d_1 U - d_2 V - d_3 W.$$

Let us take $d' = \min\{d_1, d_2, d_3\}$,

we get,　　　$\dfrac{dH}{dt} \le \Lambda - d'\, H.$

Therefore,　　$\dfrac{dH}{dt} + d'\, H \le \Lambda.$

So,　　　　$0 \le H(t) \le H(0)\, e^{-d't} + \dfrac{\Lambda}{d'}.$

when t *approaches towards* ∞ then, $0 \le H \le \dfrac{\Lambda}{d'}.$

The given system is bounded.

4. Stability analysis

The given system has three possible equilibrium states:

(i) The equilibrium state $e_1 \left(\dfrac{\Lambda}{d_1}, 0, 0 \right)$ continually exists.

(ii) The natural enemy-free equilibrium state

$$e_2 \left(\dfrac{d_2}{c_1}, \dfrac{\Lambda c_1 - d_1 d_2}{d_2 h_1}, 0 \right) \text{ exists only,}$$

if $\Lambda c_1 > d_1 d_2$ holds.

(iii) The interior steady state

$$e^* \left(U^* = \dfrac{\Lambda c_2}{c_2 d_1 + d_3 h_1}, \quad V^* = \dfrac{d_3}{c_2}, \quad W^* = \dfrac{\Lambda c_1 c_2 - d_2 (c_2 d_1 + d_3 h_1)}{(c_2 d_1 + d_3 h_1) h_2} \right)$$

exists only if $\Lambda > \dfrac{d_2 (c_2 d_1 + d_3 h_1)}{c_1 c_2}$ holds.

Now, let us discuss, the local dynamical behaviour of all non-negative equilibrium states
of the system (1)-(3), with the help of Routh Hurwitz criteria [Ahmad & Rao, 1980].

5. Routh Hurwitz criteria [12,13]

5.1　　　Theorem:　　Given　　the　　polynomial,
$P(v) = v^\omega + o_1 v^{\omega-1} + \cdots + o_{\omega-1} v + o_\omega$, where the coefficients, are real
constants, define ω matrices using the coefficients o_i for the polynomial $P(v)$ as:

$$H_1 = (o_1), \quad H_2 = \begin{pmatrix} o_1 & 1 \\ o_3 & o_2 \end{pmatrix}, \quad H_3 = \begin{pmatrix} o_1 & 1 & 0 \\ o_3 & o_2 & o_1 \\ o_5 & o_4 & o_3 \end{pmatrix} \quad \text{and}$$

$$H_\omega = \begin{pmatrix} o_1 & 1 & 0 & 0 & \cdots & 0 \\ o_3 & o_2 & o_1 & 1 & \cdots & 0 \\ o_5 & o_4 & o_3 & o_2 & \cdots & 0 \\ \vdots & \vdots & \vdots & \vdots & \cdots & \vdots \\ 0 & 0 & 0 & 0 & \ddots & o_\omega \end{pmatrix}$$

where $o_j = 0$ *if* $j > \omega$. All solutions of $P(v)$ are not positive or have non-positive real part if the determinants of all above matrices are non-negative:

$|H_j| > 0, \; j = 1, 2, ---, \omega$. When $\omega = 2$, the **Routh Hurwitz** criteria

simplifies as $|H_1| = o_1 > 0$ and $|H_2| = \begin{vmatrix} o_1 & 0 \\ 0 & o_2 \end{vmatrix} = o_1 o_2 > 0$ or $o_1 > 0$

and $o_2 > 0$.

For polynomials of degree 2, 3, 4, 5, the said criteria is as:

$\omega = 2$: $o_1 > 0$ *and* $o_2 > 0$.

$\omega = 3$: $o_1 > 0, o_3 > 0$ *and* $o_1 o_2 > o_3$.

$\omega = 4$: $o_1 > 0, o_3 > 0, o_4 > 0$ *and* $o_1 o_2 o_3 > o_3^2 + o_1^2 o_4 > 0$.

$\omega = 5$: $o_i > 0, i = 1,2,3,4,5, \; o_1 o_2 o_3 > o_3^2 + o_1^2 o_4 > 0$ *and* $(o_1 o_4 - o_5)(o_1 o_2 o_3 - o_3^2 - o_1^2 o_4) > o_5 (o_1 o_2 - o_3)^2 + o_1 o_5^2$.

5.2 Theorem The local dynamical behaviour of given system is as:

(i) The constant equilibrium state e_1 is stable only, when $\Lambda c_1 < d_1 d_2$.

(ii) If $A_1 > 0, \; A_2 > 0, \; A_3 > 0$ and $A_1 A_2 - A_3 > 0$, then the natural enemy free state e_2 is locally asymptotically stable.

(iii) If $B_1 > 0, \; B_2 > 0, \; B_3 > 0$ and $B_1 B_2 - B_3 > 0$, then the interior state e^* is locally asymptotically stable.

Proof: (i) The variational matrix of the model is

$$V = \begin{bmatrix} -d_1 - Vh_1 & -Uh_1 & 0 \\ Vc_1 & Uc_1 - d_2 - Wh_2 & -Vh_2 \\ 0 & Wc_2 & Vc_2 - d_3 \end{bmatrix}$$

The variational matrix at e_1 is

$$V(e_1) = \begin{bmatrix} -d_1 & -\dfrac{\Lambda h_1}{d_1} & 0 \\ 0 & \dfrac{\Lambda c_1}{d_1} - d_2 & 0 \\ 0 & 0 & -d_3 \end{bmatrix}$$

The characteristic equation for $V(e_1)$ is $|V(e_1) - \lambda I| = 0$.

$$\Rightarrow \begin{vmatrix} -d_1 - \lambda & -\dfrac{\Lambda h_1}{d_1} & 0 \\ 0 & \dfrac{\Lambda c_1}{d_1} - d_2 - \lambda & 0 \\ 0 & 0 & -d_3 - \lambda \end{vmatrix} = 0,$$

$$\lambda^3 + \lambda^2\left(-\frac{\Lambda c_1}{d_1} + d_1 + d_2 + d_3\right) + \lambda\left(-\Lambda c_1 + d_1 d_2 - \frac{\Lambda c_1 d_3}{d_1} + d_1 d_3 + d_2 d_3\right) + \left(-\Lambda c_1 d_3 + d_1 d_2 d_3\right) = 0$$

(4)

The characteristic roots of (4) are $\lambda = -d_1, -d_3, \dfrac{\Lambda c_1}{d_1} - d_2$. It is clear

that two roots of equation (4) are negative and one root is condi-

tional negative, i.e., $\left(\dfrac{\Lambda c_1}{d_1} - d_2\right) < 0,\quad i.e.,\ \Lambda c_1 < d_1 d_2$. Thus,

steady state e_1 is stable only when $\Lambda c_1 < d_1 d_2$ holds good.

(ii) The variational matrix of the given system at

$e_2\left(\dfrac{d_2}{c_1}, \dfrac{\Lambda c_1 - d_1 d_2}{d_2 h_1}, 0\right)$ is

170

$$V(e_2) = \begin{bmatrix} -d_1 - \dfrac{\Lambda c_1 - d_1 d_2}{d_2} & -\dfrac{d_2}{c_1}h_1 & 0 \\[3mm] \dfrac{\Lambda c_1 - d_1 d_2}{d_2 h_1}c_1 & 0 & -\dfrac{\Lambda c_1 - d_1 d_2}{d_2 h_1}h_2 \\[3mm] 0 & 0 & \dfrac{\Lambda c_1 - d_1 d_2}{d_2 h_1}c_2 - d_3 \end{bmatrix}.$$

The characteristic equation for, $V(e_2)$ is $|V(e_2) - \lambda I| = 0$,

$$\Rightarrow \begin{vmatrix} -d_1 - \dfrac{\Lambda c_1 - d_1 d_2}{d_2} - \lambda & -\dfrac{d_2}{c_1}h_1 & 0 \\[3mm] \dfrac{\Lambda c_1 - d_1 d_2}{d_2 h_1}c_1 & 0 - \lambda & -\dfrac{\Lambda c_1 - d_1 d_2}{d_2 h_1}h_2 \\[3mm] 0 & 0 & \dfrac{\Lambda c_1 - d_1 d_2}{d_2 h_1}c_2 - d_3 - \lambda \end{vmatrix} = 0$$

$$\Rightarrow \lambda^3 + A_1\lambda^2 + A_2\lambda + A_3 = 0. \tag{5}$$

Where $A_1 = \left(\dfrac{\Lambda c_1}{d_2} + d_3 + \dfrac{c_2 d_1}{h_1} - \dfrac{\Lambda c_1 c_2}{d_2 h_1} \right)$,

$A_2 = \left(\Lambda c_1 - d_1 d_2 + \dfrac{\Lambda c_1 d_3}{d_2} - \dfrac{\Lambda^2 c_1^2 c_2}{d_2^2 h_1} + \dfrac{\Lambda c_1 c_2 d_1}{d_2 h_1} \right)$

and $A_3 = \left(\Lambda c_1 d_3 - d_1 d_2 d_3 + \dfrac{2\Lambda c_1 c_2 d_1}{h_1} - \dfrac{\Lambda^2 c_1^2 c_2}{d_2 h_1} - \dfrac{c_2 d_1^2 d_2}{h_1} \right)$.

Hence by **5.1 Theorem**, all the solutions of (5) have negative real parts and state e_2 is locally asymptotically stable if $A_1 > 0$, $A_2 > 0$, $A_3 > 0$ and $A_1 A_2 - A_3 > 0$ hold.

(iii) The variational matrix of the given system at $e^* \left(\dfrac{\Lambda c_2}{c_2 d_1 + d_3 h_1}, \dfrac{d_3}{c_2}, \dfrac{\Lambda c_1 c_2 - d_2 (c_2 d_1 + d_3 h_1)}{(c_2 d_1 + d_3 h_1) h_2} \right)$ is

$$V(e^*) = \begin{bmatrix} -d_1 - \dfrac{d_3 h_1}{c_2} & -\dfrac{\Lambda c_2 h_1}{c_2 d_1 + d_3 h_1} & 0 \\[4mm] \dfrac{c_1 d_3}{c_2} & -d_2 + \dfrac{\Lambda c_2 c_1}{c_2 d_1 + d_3 h_1} - \dfrac{\Lambda c_2 c_1 - d_2(c_2 d_1 + d_3 h_1)}{c_2 d_1 + d_3 h_1} & -\dfrac{d_3 h_2}{c_2} \\[4mm] 0 & \dfrac{c_2(\Lambda c_2 c_1 - d_2(c_2 d_1 + d_3 h_1))}{(c_2 d_1 + d_3 h_1) h_2} & 0 \end{bmatrix}.$$

The characteristic equation for $V(e^*)$ is $\left| V(e^*) - \lambda I \right| = 0$,

$$\Rightarrow \begin{vmatrix} -d_1 - \dfrac{d_3 h_1}{c_2} - \lambda & -\dfrac{\Lambda c_2 h_1}{c_2 d_1 + d_3 h_1} & 0 \\[4mm] \dfrac{c_1 d_3}{c_2} & -d_2 + \dfrac{\Lambda c_2 c_1}{c_2 d_1 + d_3 h_1} - \dfrac{\Lambda c_2 c_1 - d_2(c_2 d_1 + d_3 h_1)}{c_2 d_1 + d_3 h_1} - \lambda & -\dfrac{d_3 h_2}{c_2} \\[4mm] 0 & \dfrac{c_2(\Lambda c_2 c_1 - d_2(c_2 d_1 + d_3 h_1))}{(c_2 d_1 + d_3 h_1) h_2} & 0 - \lambda \end{vmatrix} = 0.$$

$$\Rightarrow \lambda^3 + B_1 \lambda^2 + B_2 \lambda + B_3 = 0.$$

(6)

Where $B_1 = \left(d_1 + d_2 + \dfrac{h_1 d_3}{c_2} - \dfrac{c_2 d_1 d_2}{c_2 d_1 + d_3 h_1} - \dfrac{d_2 d_3 h_1}{c_2 d_1 + d_3 h_1} \right)$,

$$B_2 = \left(\begin{array}{l} d_1 d_2 + \dfrac{h_1 d_2 d_3}{c_2} - \dfrac{d_2 d_1^2 c_2}{c_2 d_1 + d_3 h_1} + \dfrac{\Lambda c_1 c_2 d_3}{c_2 d_1 + d_3 h_1} - \dfrac{c_2 d_1 d_2 d_3}{c_2 d_1 + d_3 h_1} + \dfrac{\Lambda c_1 h_1 d_3}{c_2 d_1 + d_3 h_1} \\[4mm] - \dfrac{2 d_1 d_2 d_3 h_1}{c_2 d_1 + d_3 h_1} - \dfrac{d_2 d_3^2 h_1}{c_2 d_1 + d_3 h_1} - \dfrac{d_2 d_3^2 h_1^2}{c_2(c_2 d_1 + d_3 h_1)} \end{array} \right)$$

and

$$B_3 = \left(\dfrac{\Lambda c_1 c_2 d_1 d_3}{c_2 d_1 + d_3 h_1} - \dfrac{c_2 d_1^2 d_2 d_3}{c_2 d_1 + d_3 h_1} + \dfrac{\Lambda c_1 d_3^2 h_1}{c_2 d_1 + d_3 h_{11}} - \dfrac{2 d_1 d_2 d_3^2 h_1}{c_2 d_1 + d_3 h_1} - \dfrac{d_2 d_3^2 h_1^2}{c_2(c_2 d_1 + d_3 h_1)} \right).$$

Hence by **5.1 Theorem**, all the solutions of (6) have negative real parts and the state e^* is locally asymptotically stable if $B_1 > 0$, $B_2 > 0$, $B_3 > 0$ and $B_1 B_2 - B_3 > 0$ hold.

6. Sensitivity analysis

Here, the sensitivity analysis of the system (1)-(3) for model parameters e^* is presented. The forward sensitive indices the e^* are shown in Table 1, taking the parametric values as: $\Lambda = 1.5$; $h_1 = 0.1$; $d_1 = 0.01$; $c_1 = 0.03$; $h_2 = 0.6$; $d_2 = 0.2$; $c_2 = 0.05$; $d_3 = 0.03$. Clearly Λ and c_2 have a positive impact on U^*. Whereas, h_1, d_1, d_3 have negative on U^* and other parameters have nil impact on U^*. Also, Λ is more sensitive to U^*. The parameter d_3 has a positive impact on V^*; the impact of c_2 is negative on V^* and remaining parameters have zero impact on V^*. Clearly c_2, d_3 are more sensitive parameter to V^*. Also, the impact of parameters Λ, c_1, c_2 is positive on W^* and the impact of parameters h_1, d_1, h_2, d_2, d_3 on W^* is negative. The more sensitive parameters W^* are Λ, c_1.

6.1 The sensitivity indices $\gamma^{u_x}_{p_y} = \dfrac{\partial u_x}{\partial p_y} \times \dfrac{p_y}{u_x}$ of the essential variables of the model (1)-(3) to the parameters p_y having parameter values: $\Lambda = 1.5$; $h_1 = 0.1$; $d_1 = 0.01$; $c_1 = 0.03$; $h_2 = 0.6$; $d_2 = 0.2$; $c_2 = 0.05$; $d_3 = 0.03$.

Table 1

Parameter (p_y)	$\gamma^{U^*}_{p_y}$	$\gamma^{V^*}_{p_y}$	$\gamma^{W^*}_{p_y}$
Λ	1	0	1.45161
h_1	-0.857143	0	-1.24424
d_1	-0.142857	0	-0.207373
c_1	0	0	1.45161
h_2	0	0	-1
d_2	0	0	-0.451613
c_2	0.857173	-1	1.24424
d_3	-0.857173	1	-1.24424

7. Simulations

Simulations are provided to validate the analytic findings using suitable software for distinct sets of parameters as shown in table 2. The boundary equilibrium $e_1(5,0,0)$ is stable for set-1 and result is shown in figure 1. It is far found that the natural enemy-free state $e_2(66.6,0.5,0)$ is stable for set-2 and result is shown in figure 2. The coexisting state $e^*(21.4,0.6,0.74)$ is stable for set-3 and result is shown in figure 3.

7.1 Different sets of parametric values.

Table 2

Parameter (p_y)	Set-1	Set-2	Set-3
Λ	0.5	1	1.5
h_1	0.1	0.01	0.1
d_1	0.1	0.01	0.01
c_1	0.03	0.003	0.03
h_2	0.6	0.6	0.6
d_2	0.2	0.2	0.2
c_2	0.05	0.001	0.05
d_3	0.03	0.03	0.03

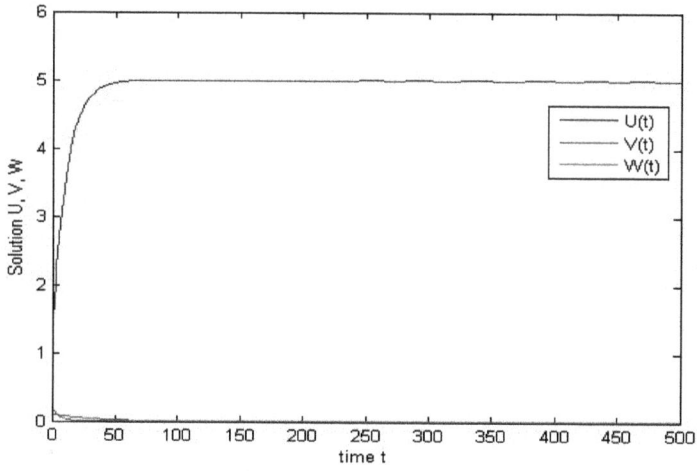

Figure 1

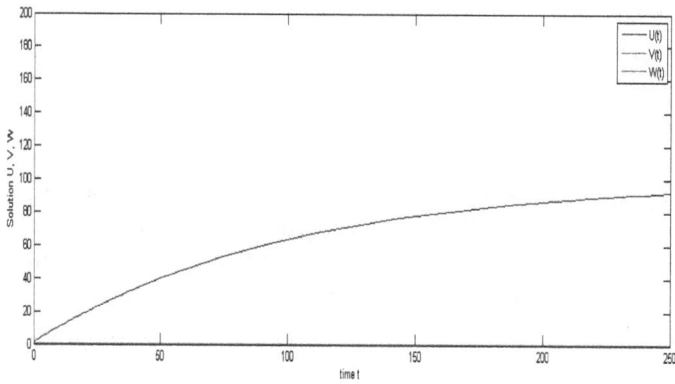

Figure 2

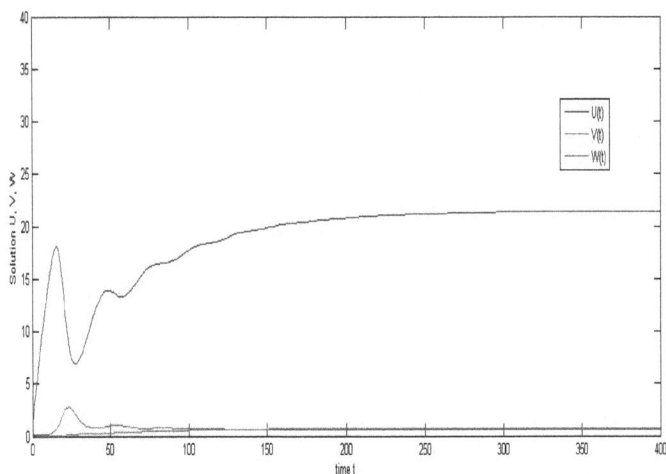

Figure 3

8. Conclusions

In this chapter, stability analysis of the crop-pest-natural enemy system as biological control approach is presented. Presented
the asymptotic stability of the system for all possible points. It is
observed that the boundary equilibrium, the natural enemy-free
equilibrium, and interior equilibrium are locally asymptotically
stable under certain conditions on parameters as discussed in 5.2
Theorem. Finally, the forward sensitivity indices are evaluated for
variables at the coexisting point for system. Simulations are
performed with a distinct set of parameters to validate our analytic
findings.

References

[1] Kishimba M., Henry L., Mwevura H., Mmochi A., Mihale M., Hellar
H. (2004): The status of pesticide pollution in tanzania. Talanta
64(1), 48–53.
[2] Weaver R.D., Evans, D.J., Luloff A.E. (1992): Pesticide use in to-
mato production: consumer concerns and willingness-to-pay. Agribusiness **8**(2), 131–142.
[3] Jain P. C., Bhargava M.C.: Entomology (2007): Novel Approaches.
New India Publishing Agency, New Delhi, India.

[4] Kar T.K., Ghosh B. (2012): Sustainability and optimal control of an exploited prey predator system through provision of alternative food to predator. BioSystems (109), 220–232.

[5] Kumar V., Dhar J., Bhatti H.S. (2020): Bifurcation dynamics of a plant-pest natural enemy system in polluted environment incorporating gestation delays. Ricerche di Matematica **69**, 533–551.

[6] Kumar V., Dhar J., Bhatti H.S. (2018): Stability and hopf-bifurcation dynamics of a food chain system: plant-pest-natural enemy with dual gestation delay as a biological control strategy. Modeling Earth Systems and Environment **4**, 881–889.

[7] Singh H., Dhar J., Bhatti H.S. (2016): Dynamics of a prey generalized predator system with disease in prey and gestation delay for predator. Modeling Earth Systems and Environment **2**(2), 52.

[8] Lian F., Xu Y. (2009): Hopf bifurcation analysis of a predator-prey system with holling type IV functional response and time delay. Applied Mathematics and Computation **215**(4), 1484–1495.

[9] Liu X., Han M. (2011): Chaos and hopf bifurcation analysis for a two species predator-prey system with prey refuge and diffusion. Nonlinear Analysis: Real World Applications **12**(2), 1047–1061.

[10] Song Y., Wei J. (2005): Local hopf bifurcation and global periodic solutions in a delayed predator-prey system. Journal of Mathematical Analysis and Applications **301**(1), 1–21.

[11] Zhao H., Lin Y. (2009): Hopf bifurcation in a partial dependent predator-prey system with delay. Chaos, Solitons and Fractals **42**(2), 896–900.

[12] Ahmad S. & Rao M.R.M. (1980): Theory of ordinary differential equations with applications in biology and engineering. East-West Press, Pvt. Ltd.

[13] Ahmad S. & Rao M.R.M. (1999): Theory of ordinary differential equations with applications in biology and engineering. Affiliated East-West Press, Pvt. Ltd.

[14] Wang W., Mulone G., Salemi F., Salone V. (2001): Permanence and stability of a stage-structured predator-prey model. Journal of Mathematical Analysis and Applications **262**(2), 499–528.

[15] Shigui R. (2001): Absolute stability, conditional stability and bifurcation in kolmogorov-type predator-prey systems with discrete delays. Quarterly of Applied Mathematics **59**(1), 159–174.

[16] Maiti A., Paul R., Alam S. (2016): A ratio-dependent predator-prey model with strong allee effect in the prey and an alternative food source for the predator. International Journal of Research in Engineering and Technology **5**(9), 233–241.

CHAPTER 11

Mathematical modeling on carrier dependent diseases with optimal control strategies

[1]Krishna Pada Das, [2]Kulbhushan Agnihotri, [3]Partha Karmakar, [4]Sanat Kumar Mahato, [5]Rakesh Kumar, [6]S. Balamuralitharan and [7]Sanjukta Pramanik
[1]Department of Mathematics, Mahadevananda Mahavidyalaya, Barrackpore, India
[2,5]Department of Mathematics, Shaheed Bhagat Singh State University, Ferozepur, Punjab, India
[3]Hoher Education Department of Govt. of West Bengal and Advisor at West Bengal Central School Service Commission, Salt Lake City, Kolkata, India
[4]Department of Mathematics, Sidho Kanho Birsha University, Purulia, India
[6]Department of Mathematics, Bharath Institute of Higher Education and Research, Chennai, Tamil Nadu, India.
[7]Chakdaha College, Chakdaha, West Bengal, India

1. Introduction

Globally, the span of ailments is a severe warning to human-race. On a large scale, the countries disburse a lump sum aggregate of their reserve on Medicare and wellness program accompaniments to treat diseased people and this straightway lays impact on the financial system pertaining to stated lands. Owing to diverse ailments, zootonic infections happen to be deliberate amplifying healthful living ultimatum across the globe. In 2015, [1] WHO gave an account of the fact the one in question, zootonic infections triggered exceeding 400,000 demise in addition to one-third of entire demises materialized in youngsters beneath a period of 5 years . As a result of deficiency of literacy, health and hygiene facilities, consciousness, the progressing countries are exceedingly concerned abreast of zootonic infections. Expansion pertaining to zootonic ailments could be regulated alongside transmitting awareness related to secured depository of food and upgrading hygiene of the environment.

Basically, houseflies, ticks, bugs and-so-forth, generally called bearers, appear to be accountant considering unfurl of zootonic infections. World Health Organization provided with a statement that there are additional 60 ailments, such as Paratyphoid fever, gastroenteritis, Hansen's disease, phthisis what ever unfurl toward people as a consequence of the existence of houseflies as well. It is noteworthy that houseflies carry away the microorganisms of infections toward consumables of people populace. Basically, flies endure mortal meal, trash, carnal faeces, and-so-forth. They can grasp their food liquified or solidified condition dissolvable in fluid secretions by salivary gland. Microbes of ailments cohere to the toe and figure of flies being shifted to the consumables of people whilst remaining flies rest on people's eatables whenever houseflies recline on domestic waste. In such manner human's edibles attain contamination. Nevertheless, no straight connection linking people populace aalongside houseflies exists yet existence of houseflies within atmosphere leads to a leap in the number of diseased human beings.

The life cycle of houseflies is generally 2-3 weeks but it might extend to three months in the admissible surroundings. The houseflies develop into quadruple phases specifically egg, maggot, pupa, adult. Subjected to the conditions of habitat (like warmth, moisture,and-so-forth), the moment grasped in evolving adult fly out-of egg has been 6 to 42 days. Having observed that denseness of houseflies becomes highest provided warmth ranges from 20^0 C to 25^0 C, nevertheless this will become exceedingly small provided warmth exceeds 45^0C or lies beneath 10^0C.WHO has endorsed that in the course of [5] eruption of cholera, dysentery, and so forth, essentially effectual reagents must be showered momentarily to sway flies as soon as required while they evolve alienation rapidly.

In order to sway the flies, chemicals like organ phosphorus compounds and pyrethroids are applied. [3] Dichloro diphenyl trichloroethane (DDT) may additionally be applied into major arenas pertaining to metropolis along-with settlements in-order-to sway flies. Since certain synthetic germicides are additionally detrimental to mortal physical state, it has been advocated, at every platform synthetic germicides ought to be applied for one month at an intermission of seven days. Prior, great deal pertaining to analysis and investigation has been executed in order to acknowledge gesture corresponding to straightway communicated ailments along-with zootonic ailments provided minor awareness has been provided to recognize the character of houseflies along-with expansion of ailment.

Flies can move considerable intervals alongside being captivated to decomposing primal matter, including mortal debris along with excretion, comprising of huge chunk of bacterial infections. [6] According to this representation, gesture pertaining to contagious ailment has been learned, taking into consideration two modes of ailment conveyance; especially (i) direct transferal and (ii) indirect transferal. It has also been reflected that bearer populace ensures logistic growth alongside magnification corresponding to growth rate carrier population. Model observation along with inspection wrapped up that number pertaining to individuals prompts growth as a result of the existence of bearer populace and people interference, whereby preferring carriers growth. The current representation has been universalized abreast by various writers in order to recognize gesture pertaining to span referring to bearer depended disease along with attainable control measures. To examine the span of ailments in existence of carriers, authors have formulated a serum (vaccine) representation along with acquiring reproduction number prompted by serum [7, 15]. Having displayed that formulated representation parades backwards bifurcation model and acquired the vaccine induced reproduction digit beneath one is by-no-means adequate for uprooting ailment. Shukla et al. (2011) contemplated gesture pertaining to contagious ailment as a result of existence appertaining to bearers along with microbes within habitat.

Through above assessment, it might be noteworthy in-order-to pause expansion pertaining to infections, such as Asiatic cholera, typhus fever and-so-forth., it seems necessary in-order-to examine bearer populace close by apartment buildings. With reference to this, Misra et al. (2012, 2013) prepared scientific models in-order-to check bearer populace employing synthetic germicides. Estimation of synthetic substances being sprinkled correlative to density pertaining to bearer populace alongwith more-or-less lag of time has been estimated in this academic work [18]. Although detainment pertaining to showering might arise unsteadiness in arrangement it has been displayed showering corresponding to synthetic germicides bears capability in-order-to sway ailment [8]. Since synthetic substance uutilization in-order-to slay bearers captivates few unfavourable aftermath on mortal fitness hence few adequate and effective tactics to lessen the denseness of houseflies along with number pertaining to people making usage corresponding to slightest volume of synthetic substances. Possessing such thoughts, a nonlinear mathematical representation has been suggested in-order-to sway houseflies reliant ailments along pertaining to synthetic substances.

2. Characterization of mathematical model

Owing to expansion, together with sway, corresponding to house-flies' reliant ailments alongside particular prominence related to zootonic and hydro-borne ailments such as gastroenteritis and-so-forth, a mathematical form has been developed by us. [4] It is under consideration that microorganisms of ailments existing within habi-tat have been carried away to mortal foodstuff as well as water by medium of houseflies. Bestrewing of synthetic germicides within the atmosphere regulates denseness pertaining to houseflies.

Presuming, within our attention belt, gross mortal populace $P(t)$, at any time t, be classified into two segments, susceptible accompanied by dimensions $S_p(t)$ as well as diseased accompanied by dimensions $I_p(t)$. Let us consider such ailment could be transferred via house-flies having density $D(t)$ as well as ceased besides swaying house-flies employing chemical germicides possessing concentration $C_p(t)$. Momentarily, the gesture possessed by houseflies' reliant ailments could be portrayed through system corresponding to succeeding non-linear ordinary differential equations:

$$dS_p / dt = \Lambda - \lambda S_p D/(m+D) - \mu S_p + \gamma I_p - \beta S_p$$
$$dI_p / dt = \lambda S_p D/(m+D) - (\gamma + \alpha + \mu)I_p + \beta S_p$$
$$dD / dt = SD\{1-(D/K)\} + S_1 HD - S_0 D - \pi_1 \theta_1 DC_p /(n+D)$$
$$dC_p/dt = \theta D/(n+D) - \theta_0 C_p - \theta_1 DC_p/(n+D) \tag{1}$$

Where $S_p(0) = S_{p0} > 0$, $I_p(0) = I_{p0} \geq 0$, $D(0) = D_0 \geq 0$, $C_p(0) = C_{p0} \geq 0$.

and where β= External source of ailment due to contaminated food;
Λ= rate of immigration of susceptible entity;
μ= Per capita natural demise rate pertaining to susceptible together with infected entity;
λ= Transmission rate of ailment from susceptible to infected indi-viduals;
m= Half saturation point at which transferral rate becomes half, while density pertaining to carrier population reaches at m;
γ= per capita recovery rate pertaining to contagious entity;
α= Per capita ailment spurred demise death rate pertaining to con-tagious entity;
s= Intrinsic growth rate pertaining to houseflies;
k= Carrying capacity pertaining to houseflies;

s_1= Growth rate pertaining to houseflies attributable to anthropogenic pursuit;

s_0= Mortality rate corresponding to houseflies attributable to components of habitat;

θ= Introduction rate pertaining to synthetic germicides in habitat of society;

θ_0= Natural depletion rate pertaining to synthetic germicides;

θ_1= Depletion rate owing to synthetic germicides attributable to its consumption by houseflies;

π_1= Proportionality constant signifying decline in growth rate of houseflies attributable to consumption pertaining to synthetic germicides.

Using $S_p+I_p= P$ within the system (1), identical model system is given by:

$$(dI_p / dt) = \lambda(P-I_p)D/(m+D)- (\gamma+\alpha+\mu)I_p + \beta(P-I_p)$$
$$(dP / dt)= \Lambda- \mu P- \alpha I_p$$
$$(dD / dt) = SD\{1-(D/K)\}+ S_1PD - S_0D -\pi_1\theta_1DC_p /(n+D)$$
$$(dC_p/dt)=\theta D/(n+D)-\theta_0C_p-\theta_1DC_p/(n+D) \qquad (2)$$

The variables engaged within aforesaid form represent positive constants. Elucidation owing to variables pertaining to (1) has been provided.

Henceforth, model system (2) has to be scrutinized elaborately.

Corresponding to boundedness pertaining to solutions regarding model system (2), basin of attraction has been stated succeeding lemma.

Lemma 1. The collection

B= $\{(I_p,\ P,\ D,\ C_p)\epsilon R^4_+ :\ 0\leq I_p\leq P\leq(\Lambda/\mu);\ 0\leq D\leq D_m;\ 0\leq C_p\leq(\theta D_m/\ \theta_0)\}$,where

$D_m= (K/S)\times\{S+(S_1\Lambda/\mu)- s_0\}$, captivates every solution commencing within interior pertaining to positive orthant.

3. Feasibility pertaining to system's equilibriums

Having set rate of change pertaining to every dynamical parameters zero, we shall attain possible equilibriums pertaining to model system (2). Our observation states that merely two equilibrium are possible in such a case; specifically (a) ailment devoid equilibrium

(DFE) $E_0(I_p', P', 0, 0)$ along with (b) endemic equilibrium(EE) $E^*(I_p^*, P^*, D^*, C_p^*)$.

Possibility pertaining to equilibrium E_0 is trivial, thus removed. Henceforth, we aim at revealing possibility pertaining to endemic equilibrium E^*.

3.1. Possibility pertaining to endemic equilibrium E^*

With respect to point of equilibrium $E^*(I_p^*, P^*, D^*, C_p^*)$, value corresponding to I_p^*, P^*, D^*, C_p^* are attained on solving succeeding equations:

$$\lambda(P-I_p)D/(m+D)-(\gamma+\alpha+\mu)I_p+\beta(P-I_p)=0, \tag{3}$$
$$\Lambda-\mu P-\alpha I_p=0, \tag{4}$$
$$S\{1-(D/K)\}+S_1P-S_0-\pi_1\theta_1C_p/(n+D)=0, \tag{5}$$
$$\theta D/(n+D)-\theta_0C_p-\theta_1DC_p/(n+D)=0. \tag{6}$$

Using equation (4) into equation (3), we achieve equation within I_p as well as D by

$$\lambda[\{\Lambda-(\alpha+\mu)I_p\}/\mu]D/(m+D)-(\gamma+\alpha+\mu)I_p+\beta[\{\Lambda-(\alpha+\mu)I_p\}/\mu]=0. \tag{7}$$

Further, using equations (4) as well as (6) within equation (5), we achieve equation within I_p and D by

$$S\{1-(D/K)\}+S_1(\Lambda-\alpha I_p)/\mu-S_0-\pi_1\theta_1\theta D/(\theta_0n+\theta_0D+\theta_1D)=0. \tag{8}$$

In accordance with possibility pertaining to endemic equilibrium, we create plot of isoclines expressed with assistance owing to equations (7) and (8).

From equation (7), our observation is:

(a)For $I_p=0$, $D=-m\beta/(\lambda+\beta)$,

(b)$I_p=(\lambda+\beta)\Lambda/\{\lambda(\alpha+\mu)+\mu(\gamma+\alpha+\mu)+\beta(\alpha+\mu)\}$

additionally

$D= -\{m\mu(\gamma+\alpha+\mu)+m\beta(\alpha+\mu)\}/\{\lambda(\alpha+\mu)+\mu(\gamma+\alpha+\mu)+\beta(\alpha+\mu)\}$ happen to be asymptotes,

(c) $(dF/dI_p)>0$.

Moreover, from equation (8), our observation is:

(a) Owing to $I_p=0$, we have obtained polynomial equation of degree two within D given below:

$h_1D^2- h_2D- h_3=0$,

Where $h_1=(S\theta_0/K)+(S\theta_1/K)$, $h_2=\theta_1\{S+(S_1\Lambda/\mu)-S_0\}+\{S\theta_0-(S\theta_0n/K)-(S_1\theta_0\Lambda/\mu)-S_0\theta_0- \pi_1\theta\theta_1\}$,

$h_3= \theta_0\{Sn+(S_1n\Lambda/\mu)-S_0n\}$,

The aforesaid equation has a unique positive root, $D=\{h_2+(h_2^2+4h_1h_3)^{1/2}\}/2h_1$,

(b) For D=0, $I_p= (\mu/S_1\alpha)\{S+(S_1\Lambda/\mu)- S_0\}$ and

(iii) $(dD/dI_p) <0$.

Eventual, in-order-to fetch positive values corresponding to P=P*, $C_p=C_p^*$, we utilize positive values owing to $I_p= I_p^*$ along with D=D* in equation (4) as well as (6) respectively. Hence, endemic equilibrium $E^*(I_h^*, P^*, D^*, C_p^*)$ is possible.

Remark1. In this section, it has been unearthed that $(dI_p^*/dm)>0$,$(dD^*/dm)>0$, $(dI_p^*/dS_1)>0$, $(dD^*/dS_1)>0$. These suggest denseness of houseflies together with aggregate of diseased rise high since half saturation constant m along with growth rate coefficient pertaining to houseflies as a consequence pertaining to mortal ventures S_1 increases. Moreover, it has been unearthed that $(dI_p^*/d\theta)<0$ as well as $(dD^*/d\theta)<0$. These suggest since introduction rate coefficient pertaining to insecticides θ multiplies, denseness of houseflies along with aggregate of diseased multiplies.

4. Local stability analysis

Steady behavior pertaining to feasible equilibrium E_0 together with E* corresponding to model system (ii) has been regulated locally. As a result, detection owing to sign of real part corresponding to every Eigen values of characteristic polynomial derived from Jacobian matrix has been carried out at that point of equilibrium. Local stability behavior corresponding to points of equilibrium E_0 along with E* have been expressed in the succeeding theorem.

Theorem 1 Ailment devoid equilibrium E_0 is unstable along with endemic equilibrium E* being locally asymptotically stable, if

$$A_3(A_1A_2-A_3)-A_1^2A_4>0. \qquad (9)$$

Proof: In order to investigate local steady behavior pertaining to E_0 together with E*, it has been found out that sign owing to real part of roots corresponding to Jacobian matrix M_0 together with M*, estimated at E_0 along with E* respectively. Jacobian matrix related to model system (2) is:

$$M=(M_{ij})_{4x4},$$

where $M_{11}= -[\{\lambda D/(m+D)\}+(\gamma+\alpha+\mu)+\beta]$, $M_{12}=[\{\lambda/(m+D)\}+\beta]$, $M_{13}=[\lambda m(P-I_p)/(m+D)^2]$, $M_{21}= -\alpha$, $M_{22}=-\mu$, $M_{32}=S_1 D$, $M_{33}=[S(1-2D/k)+S_1 P-S_0-\{\pi_1\theta_1 C_p/(n+D)\}+\{\pi_1\theta_1 DC_p/(n+D)^2\}]$, $M_{34}=-[\pi_1\theta_1 D/(n+D)]$, $M_{43}=[\{\theta n/(n+D)^2\}-\{\theta_1 nC_p/(n+D)^2\}]$ $M_{44}=-[\theta_0+\{\theta_1 D/n+D)\}]$, $M_{14}=M_{23}=M_{24}=M_{31}=M_{41}=M_{42}=0$.

By simple observation pertaining to Jacobian matrix M_0, it has been noticed that one of the characteristic values corresponding to this matrix is $(S+S_1 P-S_0)$, which is at all times positive together with another three characteristic values that are $-(\gamma+\alpha+\mu+\beta)$, $-\mu$, $-\theta_0$ are negative. Hence, equilibrium E_0 is unstable that is equilibrium E_0 is along with stable manifold locally I_p- D-C_p space along with unstable manifold locally in D- direction.

The characteristic equation for the matrix M^* is obtained as:

$\psi^4+A_1\psi^3+A_2\psi^2+A_3\psi+A_4=0$,
where$A_1=[\lambda D^*/(m+D^*)+\gamma+\alpha+2\mu+\beta+\theta_0]+[\{S/K +\theta_1/(n+D^*)\}D^*]-[(\pi_1\theta_1 D^* C_p^*)/(n+D^*)^2]$,
$A_2=[\mu\{\lambda D^*/(m+D^*)+\gamma+\alpha+\mu+\beta\}]+[\alpha\{\lambda/(m+D^*)+\beta\}]+[\mu\{SD^*/K-\pi_1\theta_1 D^* C_p^*/(n+D^*)^2\}]+[\{SD^*/K-\pi_1\theta_1 D^* C_p^*/(n+D^*)^2\}x\{\theta_0+\theta_1 D^*/(n+D^*)\}]+[\{\pi_1\theta_1 D^*/(n+D^*)\}x\{\theta n/(n+D^*)^2-\theta_1 nC_p^*/(n+D^*)^2\}]+[\{\lambda D^*/(m+D^*)+ \gamma+\alpha+\mu+\beta\}x \{SD^*/K - \pi_1\theta_1 D^* C_p^*/(n+D^*)^2\}]+[\{\lambda D^*/(m+D^*) + \gamma+\alpha+\mu+\beta\}x\{\theta_0+\theta_1 D^*/(n+D^*)\}]+[\mu\{\theta_0+\theta_1 D^*/(n+D^*)\}]$,
$A_3=[\mu\{\lambda D^*/(m+D^*) + \gamma+\alpha+\mu+\beta\}x\{SD^*/K -\pi_1\theta_1 D^* C_p^*/(n+D^*)^2\}] +[\alpha\{\lambda/(m+D^*) +\beta\}x\{SD^*/K -\pi_1\theta_1 D^* C_p^*/(n+D^*)^2\}]+[\{\lambda m(P^*-I_p^*)/(m+D^*)^2\}x\alpha S_1 D^*]+[\mu\{\lambda D^*/(m+D^*)+ \gamma+\alpha+\mu+\beta\}X\{\pi_1\theta_1 D^*/(n+D^*)\}]+[\alpha\{\lambda/(m+D^*)+\beta\}X\{\pi_1\theta_1 D^*/(n+D^*)\}]+[\alpha\{\lambda m(P^*-I_p^*)/(m+D^*)^2\}X\{\theta_0+\theta_1 D^*/(n+D^*)\}]+[\mu\{SD^*/K - \pi_1\theta_1 D^* C_p^*/(n+D^*)^2\}X\{\theta_0+\theta_1 D^*/(n+D^*)\}]+ [\mu\{\pi_1\theta_1 D^*/(n+D^*)\}X \{\theta n/(n+D^*)^2 -\theta_1 nC_p/(n+D^*)^2\}]$,
$A_4 = [\mu\{\lambda D^*/(m+D^*)+\gamma+\alpha+\mu+\beta\}x \{SD^*/K-\pi_1\theta_1 D^* C_p^*/(n+D^*)^2\}X\{\theta_0+\theta_1 D^*/(n+D^*)\}]+[\mu\{\lambda D^*/(m+D^*)+ \gamma+\alpha+\mu+\beta\}X\{\pi_1\theta_1 D^*/(n+D^*)\} X \{\theta n/(n+D^*)^2 -\theta_1 nC_p/(n+D^*)^2\}]+[\alpha\{\lambda/(m+D^*) +\beta\}X \{SD^*/K -\pi_1\theta_1 D^* C_p^*/(n+D^*)^2\}X \{\theta_0+\theta_1 D^*/(n+D^*)\}]+[\alpha\{\lambda/(m+D^*) +\beta\}X \{\pi_1\theta_1 D^*/(n+D^*)\}X\{\theta n/(n+D^*)^2 -\theta_1 nC_p/(n+D^*)^2\}]+[\alpha S_1 D^*\{\lambda m(P^*-I_p D^*)/(m+D^*)^2\}X\{\theta_0+\theta_1^*/(n+D^*)\}]$.

Herein, we observe A1, A2, A3, A4 are all positive. At the moment, by application of Routh-Hurwitz criteria for aforesaid characteristic equation, it could be told that every characteristic values corresponding to matrix M* will be either negative or negative real part provided inequality (ix) has been fulfilled.

5. Model characterization for optimal control

Owing to prior portions, it has been observed that germicides are of foremost importance in-order-to curtailing denseness of houseflies within the habitat and hence aggregate of contagious entity. In model system (2), it has been considered that ailment expands as a consequence of which, bacteria are conveyed by houseflies' out-of habitat to eatables pertaining to people populace. Hence, denseness of houseflies together with contagious individuals are claimed to be swayed within region under attention. [9] With usage of optimal control theory, this might be achieved by selecting appropriate control parameters. With respect to such aim, two control parameters in model system (ii)have been chosen; specifically (i) rate of introduction pertaining to germicides (i.e.),along with (ii) uptake rate of germicides by houseflies (such as θ_1). Variables θ along with θ_1 are symbolized by Lebesgue measurable functions $w_1(t)$ along with $w_2(t)$, respectively over bounded interval [0, T]. At the moment, problem has been reduced to that of minimizing gross cost functional W:

$$W=[N(I_p(t)+D(t))+Q_1w_1^2(t)+Q_2w_2^2(t)]dt \qquad (10)$$

Subject to
$$(dI_p / dt) = \lambda(P-I_p)D/(m+D)- (\gamma+\alpha+\mu)I_p + \beta(P-I_p)$$
$$(dP / dt)= \Lambda- \mu P- \alpha I_p$$
$$(dD / dt) = SD\{1-(D/K)\}+ S_1PD - S_0D-\pi_1w_2(t)DC_p /(n+D)$$
$$(dC_p /dt) = w_1(t)D/(n+D) - \theta_0C_h - w_2(t)DC_p/(n+D)$$

$$(11)$$

where $I_p(0)=I_{p0}\geq0$, $P(0)= P_0>0$, $D(0)=D_0\geq0$ and $C_p(0)=C_{p0}\geq0$.
J, Q_1 along with Q_2 appear to be positive weight constants that behave like balancing units owing to integrands in the cost functional W.
$N(I_p(t) +D(t))$ represents amount consumed on infection triggered by epidemic furthermore houseflies along with terms $Q_1w_1^2(t)$ and $Q_2w_2^2(t)$ signify amount engaged within initiation of germicides to-

gether with its use for diminishing denseness of houseflies. At present, our purpose is to search for [2] a pair of optimal control $(w_1^*(t), w_2^*(t)) \in V$ for minimize objective functional W for which

$$W(w_1^*(t), w_2^*(t)) = \min_{(w1^*(t), w2^*(t)) \in V} W(w_1(t), w_2(t))$$

(12)

Where control set is defined as
$$V = \{(w_1(t), w_2(t)): 0 \le w_1(t) \le w_{1max}, 0 \le w_2(t) \le w_{2max},$$
$t \in [0,T], w_1(t), w_2(t) \text{ are Lebesgue measurable}\}$

(13)

For satisfaction, we imply $w_1(t) = w_1$, $w_2(t) = w_2$.

5.1. Presence pertaining to optimal control

With reference to occurrence corresponding to optimal control pair w_1^* along with w_2^* in V which minimizes cost functional W, we present the subsequent theorem.

Theorem 2 The optimal control pair $(w_1^*, w_2^*) \in V$ pertaining to control problem (11) along with (12) exists over bounded interval [0, T].

Proof: Employing the outcomes pertaining to *Lukes (1982)*, boundedness owing to model system (11) authenticates presence of solution of control system. Hence, control parameters along with respective state variables are non-empty. The solution of model system (11) are bounded above by a linear function in the state furthermore control. The integrand within cost functional, $N(I_p(t)+D(t))+Q_1w_1^2(t)+Q_2w_2^2(t) \le N_1 + N_2(Iw_1I^2 + Iw_2I^2)^{q/2}$, where N_1 depends on upper bounds of $I_p(t)$ along with $D(t)$. Moreover, $N_2 = \max(Q_1, Q_2)$.

5.2. Characterization of optimal control

At the moment so as to find out necessary condition of optimal control pair (w_1^*, w_2^*), Pontryagin's Maximum Principle has been employed.

Theorem 3 [11] The optimal control pair $(w_1^*, w_2^*) \in V$ with regards to system (11) which minimizes objective function (x) on bounded interval [0,T] is identified as
$$w_1^* = \max[0, \min\{(-\zeta_4 D/ 2Q_1(n+D)), w_{1max}\}]$$ (14)
$$w_2^* = \max[0, \min\{(\pi_1\zeta_3+\zeta_4) DC_p/2Q_2(n+D), w_{2max}\}]$$ (15)

Proof: [16] Pontryagin's Maximum Principle sports intricate character to transform problem of minimizing cost functional conditioned to state variables into minimizing Hamiltonian. Thus by utilizing this principle, necessary condition for optimal control along with corresponding states have been derived. Hence, Hamiltonian J is given by

$J(I_p, N, D, C_p, w_1, w_2, \zeta_1, \zeta_2, \zeta_3, \zeta_4) = N(I_h+D)+Q_1w_1^2+Q_2w_2^2+\zeta_1[\lambda(P-I_p)D/(m+D)- (\gamma+\alpha+\mu)I_p + \beta(P-I_p)]+\zeta_2[\Lambda- \mu P- \alpha I_p]+\zeta_3[SD\{1- (D/K)\} + S_1PD - S_0D -\pi_1w_2DC_p /(n+D)]+\zeta_4[w_1D/(n+D) - \theta_0 C_p - w_2DC_p /(n+D)]$

Where ζ_i (i= 1(1)4) are called adjoint variables related with their respective states with transversality conditions $\zeta_1(T)=0$, $\zeta_2(T)=0$, $\zeta_3(T)=0$ along with $\zeta_4(T)=0$. The objective functional does not depend on the states at final time as a result transversality conditions are zero. We have

$$w_1^* = -\zeta_4D/2Q_1(n+D) \quad \text{and} \quad w_2^* = (\pi_1\zeta_3+\zeta_4)DC_p/2Q_2(n+D).$$

At the moment from above discoveries together with characteristics of control set V, we get

$w_1^* = \{ 0, \text{ if } -\zeta_4D/2Q_1(n+D)<0 ; -\zeta_4D/2Q_1(n+D), \text{ if } 0\leq-\zeta_4D/2Q_1(n+D)\leq w_{1max} ; w_{1max,} \text{ if } -\zeta_4D/2Q_1(n+D)>w_{1max} \}$

$w_2^* = \{0, \text{if} (\pi_1\zeta_3+\zeta_4)DC_p/2Q_2(n+D)<0; (\pi_1\zeta_3+\zeta_4)DC_p/2Q_2(n+D), \text{ if } 0\leq(\pi_1\zeta_3+\zeta_4)DC_p/2Q_2(n+D)\leq w_{2max} ; w_{2max,} \text{if}(\pi_1\zeta_3+\zeta_4)DC_p/2Q_2(n+D)>w_{2max} \}$.

Which can be equivalently written as (14) and (15)?

5.3. Optimal system

Making utilization of optimal control functions w_1^* along with w_2^*, we are capable of writing subsequent optimality system

$dI_p / dt = \lambda(P-I_p)D/(m+D)- (\gamma+\alpha+\mu)I_p + \beta(P-I_p)$

$dP / dt= \Lambda- \mu P- \alpha I_p$

$dD / dt = SD\{1-(D/K)\}+ S_1PD - S_0D -\pi_1w_2^*DC_p /(n+D)$

$dC_p/dt=w_1^*D/(n+D) - \theta_0C_p - w_2^*DC_p/(n+D)$ (16)

with initial condition $I_p(0)=I_{p0}\geq0$, $P(0)= P_0>0$, $D(0)=D_0\geq0$ along with $C_p(0)=C_{p0}\geq0$. The corresponding adjoint system is

$(d\zeta_1/dt)=-N + \zeta_1[\lambda D/(m+D)- (\gamma+\alpha+\mu) + \beta] +\zeta_2\alpha,$

$(d\zeta_2/dt)= -\zeta_1[\lambda D/(m+D)+ \beta]+ \zeta_2\mu- \zeta_3S_1D,$

$(d\zeta_3/dt)= -N-\zeta_1[\lambda(P-I_p)m/(m+D)^2]-\zeta_3[S\{1-(2D/K)\}+ S_1P - S_0 - \pi_1w_2^*nC_p /(n+D)^2] -\zeta_4[w_1^*n/(n+D)^2 - \theta_0 C_p - w_2^*nC_p/(n+D)^2],$

$(d\zeta_4/dt)= \zeta_3[\pi_1w_2^*D/(n+D)]+ \zeta_4[\theta_0 + w_2^*D/(n+D)],$

with transversality conditions $\zeta_1(T)=0$, $\zeta_2(T)=0$, $\zeta_3(T)=0$ along with $\zeta_4(T)=0$ and w1* together with w2* have been provided in (14) along with (15).

6. Conclusion

In these times, zootonic infections are severe warning for the progressing countries also they basically expand as a result of the presence of houseflies. The usage of germicides in-order-to diminish denseness of houseflies has been one of the most intricate methods in-order-to look at the predominance of these ailments. Here, we have formulated a mathematical model for evaluation of impression of germicides on the sway of houseflies' reliant ailments by considering that microorganisms of ailment are transmitted towards eeatables of people populace via houseflies. The proposed model comprises of two equilibriums; specifically (a) disease devoid equilibrium (DFE) E0, along with (b) endemic equilibrium (EE) E*. Local stability behaviorur of both equilibrium points have been set up. It has been observed that an increase in coefficient of growth pertaining to denseness of houseflies as a result of anthropogenic activities the number of contagious individuals increases; nonetheless an augmentation in the introduction rate of insecticides diminishes denseness of houseflies as well as the number of contagious individuals. The utilization of germicides is profitable but its surplus utility has detrimental consequences on mortal hygiene and is furthermore expensive. Hence, it is necessary to come up with through which one can control the denseness of houseflies by using minimum amount of germicides.Hence , we have transformed our suggested model into an optimal control problem by considerng that the introduction rate of germicides in the habitat and the exhaustion rate of denseness of houseflies as a result of consumption of germicides range with time rather than constant, which minimizes denseness of houseflies as well as the number of invectives and the associated value in employing germicides. Presence of the solution of the control system has been set up. Pontryagin's maximum principle is applied to attain and achieve the characterization of the optimal control paths. The optimal time profiles are acknowledged for introduction rate of germicides within habitat and the declining rate of denseness of houseflies as a consequence of consumption pertaining to germicides. It has been noticed that the reciprocation of single control through initiation of germicides by houseflies is more or less

190

financial in contrast to when control is enforced. Hence as a result of expansion of houseflies reliant ailments, the uutilization of the pair of control measures will be cost-effective.

References

[1] WHO's First Ever Global Estimates of Foodborne Diseases Find Children Under 5 Account for Almost One Third of   Deaths. http://www.who.int/mediacentre/news/releases/2015/f oodborne-disease-estimates/en/.

[2] Augusto, F.B., 2013. Optimal isolation control strategies and cost-effectiveness analysis of a two-strain avian influenza model. BioSystems 113, 155–164.

[3] Baker, W.C., Scudder, H.I., Guy, E.L., 1947. The control of houseflies by DDT sprays. Public Health Rep. 62 (17), 597–612.

[4] Collins, O.C., Govinder, K.S., 2016. Stability analysis and optimal vaccination of a waterborne disease model with multiple water sources. Natural Resource Modeling 29 (3), 426–447.

[5] Das, P., Mukherjee, D., Sarkar, A.K., 2005. The study of carrier dependent infectious disease cholera. J. Biol. Syst. 13, 233–244.

[6] Gonzalez-Guzman, J., 1989. An epidemiological model for direct and indirect transmission of typhoid fever. Math. Biosci. 96, 33–46.

[7] Keiding, J., World Health Organization, 1986. Division of Vector Biology and Control. The House-Fly: Biology and Control.   http://www.who.int/iris/handle/10665/60254.

[12] Keeling, M.J., Rohani, P., 2008. Modeling Infectious Diseases in Humans and Animals. Princeton University Press, New Jersey.

[13] Kar, T.K., Batabyal, A., 2011. Stability analysis and optimal control of an SIR epidemic model with vaccination. Biosystems 104, 127–135.

[14] Kalajdzievska, D., Li, M.Y., 2011. Modeling the effects of carriers on transmission dynamics of infectious diseases. Math. Biosci. Eng. 8 (3), 711–722.

[15] Kar, T.K., Ghorai, A., Jana, S., 2012. Dynamics of pest and its predator model with disease in the pest and optimal use of pesticide. J. Theor. Biol. 310, 187–198.

[16] Kar, T.K., Jana, S., 2013. A theoretical study on mathematical modeling of an infectious disease with application of optimal control. Bio-Systems 111, 37–50.

[17] Lukes, D.L., 1982. Differential Equations: Classical to Control. Academic Press, Edinburgh.

[18] Lenhart, S., Workman, J.T., 2007. Optimal Control Applied to Biological Models. Mathematical and Computational Biology Series, Chapman. Hall/CRC Press, London/ Boca Raton.

[19] Misra, A.K., Singh, V., 2012. A delay mathematical model for the spread and control of water-borne diseases. J. Theor. Biol. 301, 49–56.

[20] Misra, A.K., Mishra, S.N., Pathak, A.L., Misra, P., Naresh, R., 2012. Modeling the effect of time delay in controlling the carrier dependent infectious disease-Cholera. App.Math. Comp. 218, 11547–11557.

[21] Misra, A.K., Mishra, S.N., Pathak, A.L., Srivastava, P.K., Chandra, P., 2013. A mathema- tical model for the control of carrier-dependent infectious diseases with direct transmission and time delay. Chaos Solitons Fractals 57, 41–53.

[22] Misra, A.K., Gupta, A., 2016. A reaction-diffusion model for the control of cholera epi- demic. J. Biol. Syst. 23 (4), 1–26.

[23] Mukandavire, Z., Tripathi, A., Chiyaka, C., Musuka, G., Nyabadza, F., Mwambi, H.G., 2011. Modelling and analysis of the intrinsic dynamics of cholera. Differ. Equ. Dyn. Syst. 19 (3), 253–265.

[24] Naresh, R., Pandey, S., Misra, A.K., 2008. Analysis of a vaccination model for carrier dependent infectious diseases with environmental effects. Nonlinear Anal. Model. Cont. 13 (3), 331–350.

[25] Pontryagin, L.S., Boltyanskii, V.T., Gamkrelidze, R.V., Mishchenko, E.F., 1962. The Mathematical Theory of Optimal Processes. Wiley, London.

[26] Singh, S., Chandra, P., Shukla, J.B., 2003. Modeling and analysis of the spread of carrier- dependent infectious diseases with environmental effects. J. Biol. Syst. 11, 325–335.

[27] Shukla, J.B., Singh, V., Misra, A.K., 2011. Modeling the spread of an infectious disease with bacteria and carriers in the environment. Nonlinear Anal. Real World Appl. 12, 2541–2551.

[28] Singh, S., Singh, J., Shukla, J.B., 2019. Modeling and analysis of the effects of density dependent contact rates on the spread of carrier dependent infectious diseases with environmental discharges. Model. Earth Syst. Environ. 5, 21–32.